Anonymus

The Classification of the Sciences

Anonymus

The Classification of the Sciences

ISBN/EAN: 9783741140983

Manufactured in Europe, USA, Canada, Australia, Japa

Cover: Foto ©berggeist007 / pixelio.de

Manufactured and distributed by brebook publishing software
(www.brebook.com)

Anonymus

The Classification of the Sciences

CLASSIFICATION OF THE SCIENCES.

PREFACE

———

In the preface to the second edition, I have described myself as resisting the temptation to amplify, which the occasion raised. Reasons have since arisen for yielding to the desire which I then felt to add justifications of the scheme set forth.

The immediate cause for this change of resolve, has been the publication of several objections by Prof. Dain in his Logic. Permanently embodied, as these objections are, in a work intended for the use of students, they demand more attention than such as have been made in the course of ordinary criticism; since, if they remain unanswered, their prejudicial effects will be more continuous.

While to dispose of these I seize the opportunity afforded by a break in my ordinary work, I have thought it well at the same time to strengthen my own argument, by a re-statement from a changed point of view.

Feb., 1871.

PREFACE

TO THE SECOND EDITION.

———

THE first edition of this Essay is not yet out of print. But a proposal to translate it into French having been made by Professor Rétboré, I have decided to prepare a new edition free from the imperfections which criticism and further thought have disclosed, rather than allow these imperfections to be reproduced.

The occasion has almost tempted me into some amplification. Further arguments against the classification of M. Comte, and further arguments in support of the classification here set forth, have pleaded for utterance. But reconsideration has convinced me that it is both needless and useless to say more —needless because those who are not committed will think the case sufficiently strong as it stands, and useless because to those who are committed additional reasons will seem as inadequate as the original ones.

This last conclusion is thrust on me by seeing how little M. Littré, the leading expositor of M. Comte, is influenced by fundamental objections the force of which he admits. After quoting one of these, he says, with a candour equally rare and admirable, that he has vainly searched M. Comte's works and his own mind for an answer. Nevertheless, he adds— "j'ai réussi, je crois, à écarter l'attaque de M. Herbert Spencer, et à sauver le fond par des sacrifices indispensables mais accessoires." The sacrifices are these. He abandons M. Comte's division of Inorganic Science into Celestial Physics and Terrestrial Physics—a division which, in M. Comte's scheme, takes precedence of all the rest; and he admits that neither logically nor historically does Astronomy come before Physics, as M. Comte alleges. After making these sacrifices, which most will think too lightly described as "sacrifices indispensables mais accessoires," M. Littré proceeds to rehabilitate the Comtean classification in a way which he considers satisfactory, but which I do not understand. In short, the proof of these incongruities affects his faith in the Positivist theory of the sciences, no more than the faith of a Christian is affected by proof that the Gospels contradict one another.

Here in England I have seen no attempt to meet the criticisms with which M. Littré thus deals. There has been no reply to the allegation, based on examples, that the several sciences do not develop in the order of their decreasing generality ; nor to the allegation, based on M. Comte's own admissions, that within each science the progress is not, as he says it is, from the general to the special ; nor to the allegation that the seeming historical precedence of Astronomy over Physics in M. Comte's pages, is based on a verbal ambiguity—a mere sloight of words; nor to the allegation, abundantly illustrated, that a progression in an ordre the reverse of that asserted by M. Comte may be as well substantiated; nor to various minor allegations equally irreconcileable with his scheme. I have met with nothing more than iteration of the statement that the sciences *do* conform, logically and historically, to the ordcr in which M. Comte places them; regardless of the assigned evidence that they *do not*.

Under these circumstances it is unnecessary for me to say more; and I think I am warranted in continuing to hold that the Comtean classification of the sciences is demonstrably untenable.

While, however, I have not entered further into the controversy, as I thought of doing, I have added at the close an already-published discussion, no longer easily accessible, which indirectly enforces the general argument.

LONDON, 23RD APRIL, 1869.

CLASSIFICATION OF THE SCIENCES.

In an essay on "The Genesis of Science," originally
published in 1854, I endeavoured to show that the
Sciences cannot be rationally arranged in serial order.
Proof was given that neither the succession in which
the Sciences are placed by M. Comte (to a criticism of
whose scheme the essay was in part devoted), nor any
other succession in which the Sciences can be placed,
represents either their logical dependence or their historical dependence. To the question—How may their
relations be rightly expressed? I did not then attempt
any answer. This question I propose now to consider.

A true classification includes in each class, those
objects which have more characteristics in common
with one another, than any of them have in common
with any objects excluded from the class. Further,
the characteristics possessed in common by the colligated objects, and not possessed by other objects, are
more radical than any characteristics possessed in
common with other objects—involve more numerous

4

dependent characteristics. These are two sides of the
same definition. For things possessing the greatest
number of attributes in common, are things that possess in common those essential attributes on which the
rest depend; and, conversely, the possession in common of the essential attributes, implies the possession
in common of the greatest number of attributes. Hence,
either test may be used as convenience dictates.

If, then, the Sciences admit of classification at all, it
must be by grouping together the like and separating
the unlike, as thus defined. Let us proceed to do this.

The broadest natural division among the Sciences,
is the division between those which deal with the abstract relations under which phenomena are presented
to us, and those which deal with the phenomena themselves. Relations of whatever orders, are nearer akin
to one another than they are to any objects. Objects
of whatever orders, are nearer akin to one another
than they are to any relations. Whether, as some
hold, Space and Time are forms of Thought; or
whether, as I hold myself, they are forms of Things,
that have become forms of Thought through organized and inherited experience of Things; it is equally
true that Space and Time are contrasted absolutely
with the existences disclosed to us in Space and Time;
and that the Sciences which deal exclusively with
Space and Time, are separated by the profoundest of
all distinctions from the Sciences which deal with the

existences that Space and Time contain. Space is the abstract of all relations of co-existence. Time is the abstract of all relations of sequence. And dealing as they do entirely with relations of co-existence and sequence, in their general or special forms, Logic and Mathematics form a class of the Sciences more widely unlike the rest, than any of the rest can be from one another.

The Sciences which deal with existences themselves, instead of the blank forms in which existences are presented to us, admit of a sub-division less profound than the division above made, but more profound than any of the.divisions among the Sciences individually considered. They fall into two classes, having quite different aspects, aims, and methods. Every phenomenon is more or less composite—is a manifestation of force under several distinct modes. Hence result two objects of inquiry. We may study the component modes of force separately ; or we may study them in their relations, as co-operative factors in this composite phenomenon. On the one hand, neglecting all the incidents of particular cases, we may aim to educe the laws of each mode of force, when it is uninterfered with. On the other hand, the incidents of the particular case being given, we may seek to interpret the entire phenomenon, as a product of the several forces simultaneously in action. The truths reached through the first kind of inquiry, though concrete inasmuch as they have actual existences for their subject-matters,

6

are abstract inasmuch as they refer to the modes of existence apart from one another; while the truths reached by the second kind of inquiry are properly concrete, inasmuch as they formulate the facts in their combined order, as they occur in Nature.

The Sciences, then, in their main divisions, stand thus:—

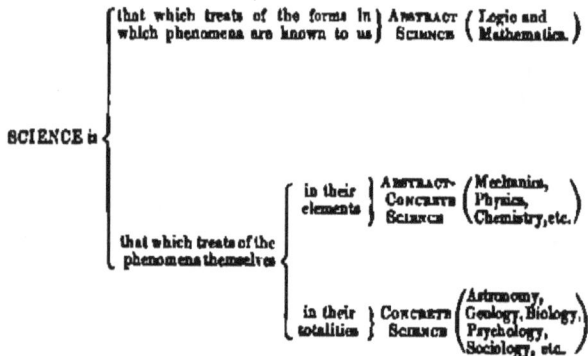

SCIENCE is

- that which treats of the forms in which phenomena are known to us} ABSTRACT SCIENCE (Logic and Mathematics.)
- that which treats of the phenomena themselves
  - in their elements } ABSTRACT-CONCRETE SCIENCE (Mechanics, Physics, Chemistry, etc.)
  - in their totalities } CONCRETE SCIENCE (Astronomy, Geology, Biology, Psychology, Sociology, etc.)

It is needful to define the words *abstract* and *concrete* as thus used; since they are sometimes used with other meanings. M. Comte divides Science into abstract and concrete; but the divisions which he distinguishes by these names are quite unlike those above made. Instead of regarding some Sciences as wholly abstract, and others as wholly concrete, he regards each Science as having an abstract part, and a concrete part. There is, according to him, an abstract mathematics and a concrete mathematics—an

abstract biology and concrete biology. He says:—
"Il faut distinguer, par rapport à tous les ordres de
phénomènes, deux genres de sciences naturelles : les
unes abstraites, générales, ont pour objet la découverte
des lois qui régissent les diverses classes de phénomènes, en considérant tous les cas qu'on peut con-
cevior ; les autres concrètes, particulières, descriptives,
et qu'on désigne quelquefois sous le nom de sciences
naturelles proprement dites, consistent dans l'application de ces lois a l'histoire effective de différons êtres
existans." And to illustrate the distinction, he names
general physiology as abstract, and zoology and botany
as concrete. Here it is manifest that the words
abstract and *general* are used as synonymous. They
have, however, different meanings ; and confusion
results from not distinguishing their meanings. Abstractness means *detachment from* the incidents of particular cases. Generality means *manifestation in* numerous
cases. On the one hand, the essential nature of some
phenomenon is considered, apart from disguising phenomena. On the other hand, the frequency of the
phenomenon, with or without disguising phenomena,
is the thing considered. Among the ideal relations of
numbers the two coincide ; but excluding these, an
abstract truth is not realizable to perception in any
case of which it is asserted, whereas a general truth is
realizable to perception in every case of which it is
asserted. Some illustrations will make the distinction
clear. Thus it is an abstract truth that the angle contained

in a semi-circle is a right angle—abstract in the sense
that though it does not hold in actually-constructed
semi-circles and angles, which are always inexact, it
holds in the ideal semi-circles and angles abstracted
from real ones; but this is not a general truth, either
in the sense that it is commonly manifested in Nature,
or in the sense that it is a space-relation that compre-
hends many minor space-relations : it is a quite
special space-relation. Again, that the momentum
of a body causes it to move in a straight line at a
uniform velocity, is an abstract-concrete truth—a
truth abstracted from certain experiences of concrete
phenomena; but it is by no means a general truth:
so little generality has it, that no one fact in Nature
displays it. Conversely, surrounding things supply
us with hosts of general truths that are not in the
least abstract. It is a general truth that the planets
go round the Sun from West to East—a truth which
holds good in something like a hundred cases (including. the cases of the planetoids); but this truth
is not at all abstract, since it is perfectly realized
as a concrete fact in every one of these cases. Every
vertebrate animal whatever, has a double nervous
system; all birds and all mammals are warm-
blooded—these are general truths, but they are
concrete truths : that is to say, every vertebrate
animal individually presents an entire and unqualified
manifestation of this duality of the nervous system;
every living bird exemplifies absolutely or completely

the warm-bloodedness of birds. What we here call, and rightly call, a general truth, is simply a proposition which *sums up* a number of our actual experiences; and not the expression of a truth *drawn from* our actual experiences, but never presented to us in any of them. In other words, a general truth colligates a number of particular truths; while an abstract truth colligates no particular truths, but formulates a truth which certain phenomena all involve, though it is actually seen in none of them.

Limiting the words to their proper meanings as thus defined, it becomes manifest that the three classes of Sciences above separated, are not distinguishable at all by differences in their degrees of generality. They are all equally general; or rather they are all, considered as groups, universal. Every object whatever presents at once the subject-matter for each of them. In the smallest particle of substance we have simultaneously illustrated the abstract truths of relation in Time and Space; the abstract-concrete truths in conformity with which the particle manifests its several modes of force; and the concrete truths which are the laws of the joint manifestation of these modes of force. Thus these three classes of Sciences severally formulate different, but co-extensive, classes of facts. Within each group there are truths of greater and less generality: there are general abstract truths, and special abstract truths; general abstract-concrete truths, and special abstract-concrete truths;

general concrete truths, and special concrete truths. But while within each class there are groups and sub-groups and sub-sub-groups which differ in their degrees of generality, the classes themselves differ only in their degrees of abstractness.*

* Some propositions laid down by M. Littré, in his lately-published book— *Auguste Comte et la Philosophie Positive*, may fitly be dealt with here. In the candid and courteous reply he makes to my strictures on the Comtean classification in "The Genesis of Science," he endeavours to clear up some of the inconsistencies I pointed out; and he does this by drawing a distinction between objective generality and subjective generality. He says—"qu'il existe deux ordres de généralité, l'une objective et dans les choses, l'autre subjective, abstraite et dans l'esprit." This sentence, in which M. Littré makes subjective generality synonymous with abstractness, led me at first to conclude that he had in view the same distinction as that which I have above explained between generality and abstractness. On re-reading the paragraph, however, I found this was not the case. In a previous sentence he says—"La biologie a passé de la considération des organes à celles des tissus, plus généraux que les organes, et de la considération des tissus à celle des éléments anatomiques, plus généraux que les tissus. Mais cette généralité croissante est subjective non objective, abstraite non concrète." Here it is manifest that abstract and concrete, are used in senses analogous to those in which they are used by M. Comte; who, as we have seen, regards general physiology as abstract and zoology and botany as concrete. And it is further manifest that the word abstract, as thus used, is not used in its proper sense. For, as above shown, no such facts as those of anatomical structure can be abstract facts; but can only be more or less general facts. Nor do I understand M. Littré's point of view when he regards those more general facts of anatomical structure, as *subjectively* general and not *objectively* general. The structural phenomena presented by any tissue, such as mucous membrane, are more general than the phenomena presented by any of the organs which mucous membrane goes to form, simply in the sense that the phenomena peculiar to the membrane are repeated in a greater number of instances than the phenomena peculiar to any organ into the composition of which the membrane enters. And, similarly, such facts as have been established respecting the anatomical elements of tissues, are more general than the facts established respecting any particular tissue, in the sense that they are facts which organic bodies exhibit in a greater number of cases—they are *objectively* more general; and they can be called *subjectively* more general only in the sense that the conception corresponds with the phenomena.

Let me endeavour to clear up this point:—There is, as M. Littré truly says, a decreasing generality that is objective. If we omit the phenomena of Dissolution, which are changes from the special to the general, all changes which matter undergoes are from the general to the special—are changes involving a decreasing

Passing to the sub-divisions of these classes, we find that the first class is separable into two parts—the one containing universal truths, the other non-universal truths. Dealing wholly with relations apart from related things, Abstract Science considers first, that which is common to all relations whatever; and second, that which is common to each order of relations. Besides the indefinite and variable connexions which exist among phenomena, as occurring together in Space and Time, we find that there are also definite

generality in the united groups of attributes. This is the progress of *things*. The progress of *thought*, is not only in the same direction, but also in the opposite direction. The investigation of Nature discloses an increasing number of specialities; but it simultaneously discloses more and more the generalities within which those specialities fall. Take a case. Zoology, while it goes on multiplying the number of its species, and getting a more complete knowledge of each species (decreasing generality); also goes on discovering the common characters by which species are united into larger groups (increasing generality). Both these are subjective processes; and in this case, both orders of truths reached are concrete—formulate the phenomena as actually manifested.

M. Littré, recognising the necessity for some modification, of the hierarchy of the Sciences, as enunciated by M. Comte, still regards it as substantially true; and for proof of its validity, he appeals mainly to the essential *constitutions* of the Sciences. It is unnecessary for me here to meet, in detail, the arguments by which he supports the proposition, that the essential constitutions of the Sciences, justify the order in which M. Comte places them. It will suffice to refer to the foregoing pages, and to the pages which are to follow, as containing the definitions of those fundamental characteristics which demand the grouping of the Sciences in the way pointed out. As already shown, and as will be shown still more clearly by and bye, the radical differences of constitution among the Sciences, necessitate the colligation of them into the three classes—Abstract, Abstract-Concrete, and Concrete. How irreconcilable is M. Comte's classification with these groups, will be at once apparent on inspection. It stands thus :—

Mathematics (including rational Mechanics), ... partly Abstract, partly Abstract-Concrete.

Astronomy .... Concrete.

Physics... Abstract-Concrete.

Chemistry .... Abstract-Concrete.

Biology... Concrete.

Sociology .... Concrete.

and invariable connexions—that between each kind of
phenomenon and certain other kinds of phenomena,
there exist uniform relations. This is a universal
abstract truth—that there is an unchanging order
among things in Space and Time. We come next
to the several kinds of unchanging order, which,
taken together, form the subjects of the second
division of Abstract Science. Of this second division, the most general sub-division is that which
deals with the natures of the connexions in Space
and Time, irrespective of the terms connected. The
conditions under which we may predicate a relation of coincidence or proximity in Space and
Time (or of non-coincidence or non-proximity) form
the subject-matter of Logic. Here the natures and
amounts of the terms between which the relations are
asserted (or denied) are of no moment : the propositions of Logic are independent of any qualitative
or quantitative specification of the related things.
The other sub-division has for its subject-matter, the
relations between terms which are specified quanti-
tatively but not qualitatively. The amounts of the
related terms, irrespective of their natures, are here
dealt with; and Mathematics is a statement of the
laws of quantity considered apart from reality. Quantity considered apart from reality, is occupancy of
Space or Time; and occupancy of Space or Time
is measured by the number of coexistent or sequent
positions occupied. That is to say, quantities can be

compared and the relations between them established, only by some direct or indirect enumeration of their component units; and the ultimate units into which all others are decomposable, are such occupied positions in Space as can, by making impressions on consciousness, produce occupied positions in Time. Among units that are unspecified in their natures (extensive, protensive, or intensive), but are ideally endowed with existence considered apart from attributes, the quantitative relations that arise, are those most general relations expressed by numbers. Such relations fall into either of two orders, according as the units are considered simply as capable of filling separate places in consciousness, or according as they are considered as filling places that are not only separate, but equal. In the one case, we have that indefinite calculus by which numbers of abstract existences, but not sums of abstract existence, are predicable. In the other case, we have that definite calculus by which both numbers of abstract existences and sums of abstract existence are predicable. Next comes that division of Mathematics which deals with the quantitative relations of magnitudes (or aggregates of units) considered as coexistent, or as occupying Space—the division called Geometry. And then we arrive at relations, the terms of which include both quantities of Time and quantities of Space—those in which times are estimated by the units of space traversed at a uniform velocity, and those in which equal

units of time being given, tho spaces traversed with
uniform or variable velocities are estimated. These
Abstract Sciences, which are concerned exclusively
with relations and with the relations of relations, may
be grouped as shown in Table I.

Passing from the Sciences that treat of the ideal or
unoccupied forms of relations, and turning to the
Sciences that treat of real relations, or the relations
among realities, we come first to those Sciences which
deal with realities, not as they are habitually manifested to us, but with realities as manifested in their
different modes, when these are artificially separated
from one another. In the same way that tho Abstract
Sciences are ideal, relatively to the Abstract-Concrete
and Concrete Sciences ; so the Abstract-Concrete
Sciences are ideal, relatively to the Concrete Sciences.
Just as Logic and Mathematics have for their object
to generalize the laws of relation, qualitative and
quantitative, apart from related things; so, Mechanics, Physics, Chemistry, etc., have for their object
to generalize the laws of relation which different
modes of Matter and Motion conform to, when severally disentangled from those actual phenomena in
which they are mutually modified. Just as the
geometrician formulates the properties of lines and
surfaces, independently of the irregularities and thicknesses of lines and surfaces as they really exist; so,
the physicist and the chemist formulate the mani-

Law of relation—an expression of the truth that uniformities of connexion obtain among modes of Being, irrespective of any specification of the of the uniformities of connexion.

Relations

that are qualitative; or that are specified in their nature as relations of coincidence or proximity in Time and Space, but not necessarily in their terms: the natures and amount of which are indifferent. (Logic.)*

that are quantitative (MATHEMATICS)

negatively: the terms of the relations being definitely-related sets of positions in space; and the facts predicated being units that are equal only as having independent existence. (Geometry of Position.**)

positively: the terms being magnitudes composed of

units that are equal only as having independent existence. (Indefinite Calculus.†)

equal units

the equality of which is not defined as extensive, or intensive, protensive, or (Definite Calculus)

the equality of which is that of extension

when their numbers are completely specified. (Arithmetic.)

when their numbers are specified only

in their relations. (Algebra.)

in the relations of their relations. (Calculus of Operations.)

considered in their relations of coexistence. (Geometry.)

considered as traversed in Time

that is wholly indefinite. (Kinematics.)

that is divided into equal units. (Geometry of Motion.‡)

*... includes the laws of re-
... geometry, but not those of
... century. There had, in
... bility of an inferred con-
h the number of times each
... curred in experience, are
... mathematically.

† of explanation of the term negatively-quantitative, it
... uses the proposition that certain three lines will meet
... ratively-quantitative proposition; since it asserts the
... ality of space between their intersections. Similarly,
... certain three points will always fall in a straight line
... m of any lateral quantity, or deviation.

... ning of this division should not be misunderstood. It may be well to
... s, the estimates of the statistician. Calculations respecting popu-
... n, etc., have results which are correct only numerically, and not
... nalities of being or action represented by the numbers.

ll be asked—How can there be a Geometry of Motion into which the com-
... ics not enter? The reply is, that the time-relations and space-relations of
... sidered apart from those of Force, in the same way that the space-relations
... considered apart from Matter.

festations of each mode of force, independently of
the disturbances in its manifestations which other
modes of force cause in every actual case. In works
on Mechanics, the laws of motion are expressed without reference to friction and resistance of the medium.
Not what motion ever really is, but what it would
be if retarding forces were absent, is asserted. If any
retarding force is taken into account, then the effect
of this retarding force is alone contemplated : neglecting the other retarding forces. Consider, again, the
generalizations of the physicist respecting molecular
motion. The law that light varies inversely as the
square of the distance, is absolutely true only
when the radiation goes on from a point without
dimensions, which it never does; and it also assumes
that the rays are perfectly straight, which they cannot
be unless the medium differs from all actual media in .
being perfectly homogeneous. If the disturbing
effects of changes of media are investigated, the
formulæ expressing the refractions take for granted
that the new media entered are homogeneous; which
they never really are. Even when a compound
disturbance is allowed for, as when the refraction
undergone by light in traversing a medium of increasing density, like the atmosphere, is calculated,
the calculation still supposes conditions that are un-
naturally simple—it supposes that the atmosphere
is not pervaded by heterogeneous currents, which
it always is. Similarly with the inquiries of the

chemist. He does not take his substances as Nature
supplies them. Before he proceeds to specify their
respective properties, he purifies them—separates from
each all trace of every other. Before ascertaining the
specific gravity of a gas, he has to free this gas from
the vapour of water, usually mixed with it. Before
describing the properties of a salt, he guards against
any error that may arise from the presence of an
uncombined portion of the acid or base. And when
he alleges of any element that it has a certain atomic
weight, and unites with such and such equivalents
of other elements, he does not mean that the results
thus expressed are exactly the results of any one
experiment; but that they are the results which,
after averaging many trials, he concludes would be
realized if absolute purity could be obtained, and
. if the experiments could be conducted without
loss. His problem is to ascertain the laws of
combination of molecules, not as they are actually
displayed, but as they would be displayed in the
absence of those minute interferences which cannot
be altogether avoided. Thus all these Abstract-Concrete Sciences have for their object, *analytical interpretation.* In every case it is the aim to decompose
the phenomenon, and formulate its components apart
from one another; or some two or three apart from
the rest. Wherever, throughout these Sciences, synthesis is employed, it is for the verification of analysis.*

* I am indebted to Prof. Frankland for reminding me of an objection that may be

The truths elaborated are severally asserted, not as
truths exhibited by this or that particular object; but
as truths universally holding of Matter and Motion in
their more general or more special forms, considered
apart from particular objects, and particular places in
space.

The sub-divisions of this group of Sciences, may be
drawn on the same principle as that on which the
sub-divisions of the preceding group were drawn.
Phenomena, considered as more or less involved
manifestations of force, yield on analysis, certain
laws of manifestation that are universal, and other
laws of manifestation, which, being dependent on
conditions, are not universal. Hence the Abstract-
Concrete Sciences are primarily divisible into—the
laws of force considered apart from its separate modes,
and laws of force considered under each of its separate modes. And this second division of the Abstract-
Concrete group, is sub-divisible after a manner essentially analogous. It is needless to occupy space by

made to this statement. The production of new compounds by synthesis, has of
late become an important branch of chemistry. According to certain known laws
of composition, complex substances, which never before existed, are formed, and
fulfil anticipations both as to their general properties and as to the proportions of
their constituents—as proved by analysis. Here it may be said with truth, that
analysis is used to verify synthesis. Nevertheless, the exception to the above
statement is apparent only—not real. In so far as the production of new compounds is carried on merely for the obtainment of such new compounds, it is not
Science but Art—the application of pre-established knowledge to the achievement
of ends. The proceeding is a part of Science, only in so far as it is a means to
the better interpretation of the order of Nature. And how does it aid the interpretation ? It does it only by verifying the pre-established conclusions respecting
the laws of molecular combination; or by serving further to explain them. That
is to say, these syntheses, considered on their scientific side, have simply the purpose of *forwarding the analysis of the laws of chemical combination*.

defining these several orders and genera of Sciences.
Table II. will sufficiently explain their relations.

We come now to the third great group. We have
done with the Sciences which are concerned only with
the blank forms of relations under which Being is
manifested to us. We have left behind the Sciences
which, dealing with Being under its universal mode,
and its several non-universal modes regarded as independent, treats the terms of its relations as simple and
homogeneous, which they never are in Nature. There
remain the Sciences which, taking these modes of
Being as they are connected with one another, have for
the terms of their relations, those heterogeneous combinations of forces that constitute actual phenomena.
The subject-matter of these Concrete-Sciences is the
real, as contrasted with the wholly or partially ideal.
It is their aim, not to separate and generalize apart
the components of all phenomena; but to explain each
phenomenon as a product of these components. Their
relations are not, like those of the simplest Abstract-
Concrete Sciences, relations between one antecedent
and one consequent; nor are they, like those of the
more involved Abstract-Concrete Sciences, relations
between some few antecedents cut off in imagination
from all others, and some few consequents similarly
cut off; but they are relations each of which has for
its terms a complete plexus of antecedents and a complete plexus of consequents. This is manifest in the

ure of forces (tensions and pressures), as deducible from the persistence of force: the theorems of resolution and composition of forces.

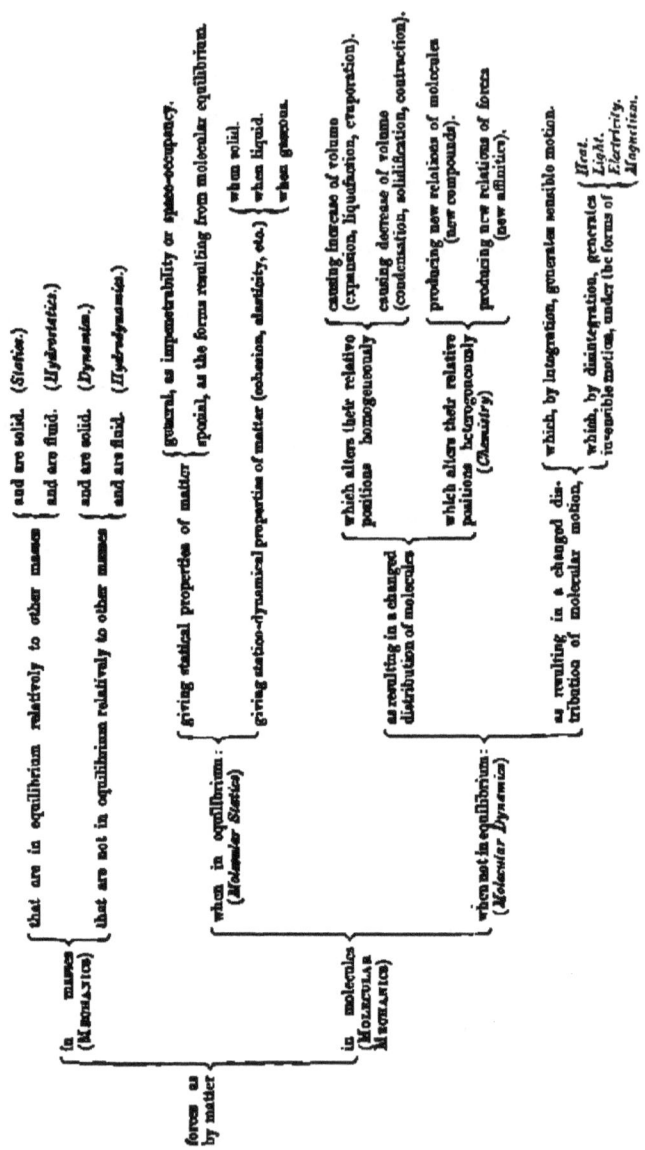

forces as by matter

in masses (MOLAR ACTION)

- that are in equilibrium relatively to other masses
  - and are solid. (*Statics.*)
  - and are fluid. (*Hydrostatics.*)
- that are not in equilibrium relatively to other masses
  - and are solid. (*Dynamics.*)
  - and are fluid. (*Hydrodynamics.*)

in molecules (MOLECULAR MOVEMENTS)

- when in equilibrium: (*Molecular Statics*)
  - giving statical properties of matter
    - general, as impenetrability or space-occupancy.
    - special, as the forms resulting from molecular equilibrium
      - when solid.
      - when liquid.
      - when gaseous.
  - giving station-dynamical properties of matter (cohesion, elasticity, etc.)
- when not in equilibrium: (*Molecular Dynamics*)
  - as resulting in a changed distribution of molecules
    - which alters their relative positions homogeneously
      - causing increase of volume (expansion, liquefaction, evaporation).
      - causing decrease of volume (condensation, solidification, contraction).
    - which alters their relative positions heterogeneously (*Chemistry*)
      - producing new relations of molecules (new compounds).
      - producing new relations of forces (new affinity).
  - as resulting in a changed distribution of molecular motion,
    - which, by integration, generates sensible motion.
    - which, by disintegration, generates irreversible motion, under the forms of
      - Heat.
      - Light.
      - Electricity.
      - Magnetism.

least involved Concrete Sciences. The astronomer
seeks to explain the Solar System. He does not stop
short after generalizing the laws of planetary movement, such as planetary movement would be did only
a single planet exist; but he solves this abstract-concrete problem, as a step towards solving the concrete
problem of the planetary movements as affecting one
another. In astronomical language, "the theory of
the Moon" means an interpretation of the Moon's
motions, not as determined simply by centripetal and
centrifugal forces, but as perpetually modified by
gravitation towards the Earth's equatorial protuber-
ance, towards the Sun, and even towards Venus—
forces daily varying in their amounts and combinations. Nor does the astronomer leave off when he has
calculated what will be the position of a given body
at a given time, allowing for all perturbing influences;
but he goes on to consider the effects produced by re-
actions on the perturbing masses. And he further
goes on to consider how these mutual perturbations
of the planets cause, during a long period, increasing
deviations from a mean state; and then how compen-
sating perturbations cause continuous decrease in the
deviations. That is, the goal towards which he ever
strives, is a complete explanation of these complex
planetary motions in their totality. Similarly with
the geologist. He does not take for his problem only
those irregularities of the Earth's crust that are
worked by denudation; or only those which igneous

action causes. He does not seek simply to understand how sedimentary strata were formed; or how faults were produced; or how moraines originated; or how the beds of Alpine lakes were scooped out. But taking into account all agencies co-operating in endless and ever-varying combinations, he aims to interpret the entire structure of the Earth's crust. If he studies separately the actions of rain, rivers, glaciers, icebergs, tides, waves, volcanoes, earthquakes, etc.; he does so that he may be better able to comprehend their joint actions as factors in geological phenomena: the object of his science being to generalize these phenomena in all their involved connections, as parts of one whole. In like manner Biology is the elaboration of a complete theory of Life, in each and all of its involved manifestations. If different aspects of its phenomena are investigated apart—if one observer busies himself in classing organisms, another in dissecting them, another in ascertaining their chemical compositions, another in studying functions, another in tracing laws of modification; they are all, consciously or unconsciously, helping to work out a solution of vital phenomena in their entirety, both as displayed by individual organisms and by organisms at large. Thus, in these Concrete Sciences, the object is the converse of that which the Abstract-Concrete Sciences propose to themselves. In the one case we have *analytical interpretation;* while in the other case we . have *synthetical interpretation.* Instead of synthesis being used merely to verify analysis; analysis is here used only to aid synthesis. Not to formulate the factors of phenomena is now the object; but to formulate the phenomena resulting from these factors, under the various conditions which the Universe presents.

This third class of Sciences, like the other classes, is divisible into the universal and the non-universal. As there are truths which hold of all phenomena in their elements; so there are truths which hold of all phenomena in their totalities. As force has certain ultimate laws common to its separate modes of manifestation, so in those combinations of its modes which constitute actual phenomena, we find certain ultimate laws that are conformed to in every case. These are the laws of the re-distribution of force. Since we can become conscious of a phenomenon only by some change wrought in us, every phenomenon necessarily implies re-distribution of force—change in the arrangements of matter and motion. Alike in molecular movements and the movements of masses, one great uniformity may be traced. A decreasing quantity of motion, sensible or insensible, always has for its concomitant an increasing aggregation of matter; and, conversely, an increasing quantity of motion, sensible or insensible, has for its concomitant a decreasing aggregation of matter. Give to the molecules of any mass, more of that insensible motion which we call heat, and the parts of the mass become somewhat less closely aggregated. Add a further quantity of insensible motion, and the mass so far disintegrates as to become liquid. Add still more insensible motion, and the mass disintegrates so completely as to become gas; which occupies a greater space with every extra quantity of insensible motion given to it. On the other hand, · every loss of insensible motion by a mass, gaseous, liquid, or solid, is accompanied by a progressing integration of the mass. Similarly with sensible motions, be the bodies moved large or small. Augment the velocities of the planets, and their orbits will enlarge—the Solar System would occupy a wider space. Diminish their velocities, and their orbits will lessen—the Solar System will contract, or become more integrated. And in like manner we see that every sensible motion on the Earth's surface involves a partial disintegration of the moving body from the Earth; while the loss of its motion is accompanied by the body's re-integration with the Earth. In all phenomena we have either an integration of matter and concomitant dissipation of motion; or an absorption of motion and concomitant disintegration of matter. And where, as in living bodies, these processes are going on simultaneously, there is an integration of matter proportioned to the dissipation of motion, and an absorption of motion proportioned to the disintegration of matter. Such, then, are the universal laws of that re-distribution of matter and motion everywhere going on—a re-distribution which results in Evolution so long as

the aggregation of matter and dispersion of motion
predominate; but which results in Dissolution where
there is a predominant aggregation of motion and
dispersion of matter. Hence we have a division
of Concrete Science which bears towards the other
Concrete Sciences, a relation like that which Universal
Law of Relation bears to Mathematics, and like that
which Universal Mechanics (composition and resolution of forces) bears to Physics. We have a division of
Concrete Science which generalizes those concomitants
of this re-distribution that hold good among all orders
of concrete objects—a division which explains why,
along with a predominating integration of matter and
dissipation of motion, there goes a change from
an indefinite, incoherent homogeneity, to a definite,
coherent heterogeneity; and why a reverse re-distribution of matter and motion, is accompanied by
a reverse structural change. Passing from this universal Concrete Science, to the non-universal Concrete
Sciences; we find that these are primarily divisible
into the science which deals with the re-distributions
of matter and motion among the masses in space, consequent on their mutual actions as wholes; and the
science which deals with the re-distributions of matter
and motion consequent on the mutual actions of the
molecules in each mass. And of these equally general
Sciences, this last is re-divisible into the Science which
is limited to the concomitants of re-distribution among
the molecules of each mass when regarded as inde-

pendent, and the Science which takes into account the molecular motion received by radiation from other masses. But these sub-divisions, and their sub-subdivisions, will be best seen in the annexed Table III.

That these great groups of Sciences and their respective sub-groups, fulfil the definition of a true classification given at the outset, is, I think, tolerably manifest. The subjects of inquiry included in each primary division, have essential attributes in common with one another, which they have not in common with any of the subjects contained in the other primary divisions; and they have, by consequence, a greater number of common attributes in which they severally agree with the colligated subjects, and disagree with the subjects otherwise colligated. Between Sciences which deal with relations apart from realities, and Sciences which deal with realities, the distinction is the widest possible; since Being, in some or all of its attributes, is common to all Sciences of the second class, and excluded from all Sciences of the first class. The distinction between the empty forms of things and the things themselves, is a distinction which cannot be exceeded in degree. · And when we divide the Sciences which treat of realities, into those which deal with their separate components and those which deal with their components as united, we make a profounder distinction than can exist between the Sciences which deal with one or other order

ternal laws of the continuous re-distribution of Matter and Motion; which results in Evolution where there is a predominant integration of Matter and dissipation f Motion, and which results in Dissolution where there is a predominant absorption of Motion and disintegration of Matter.

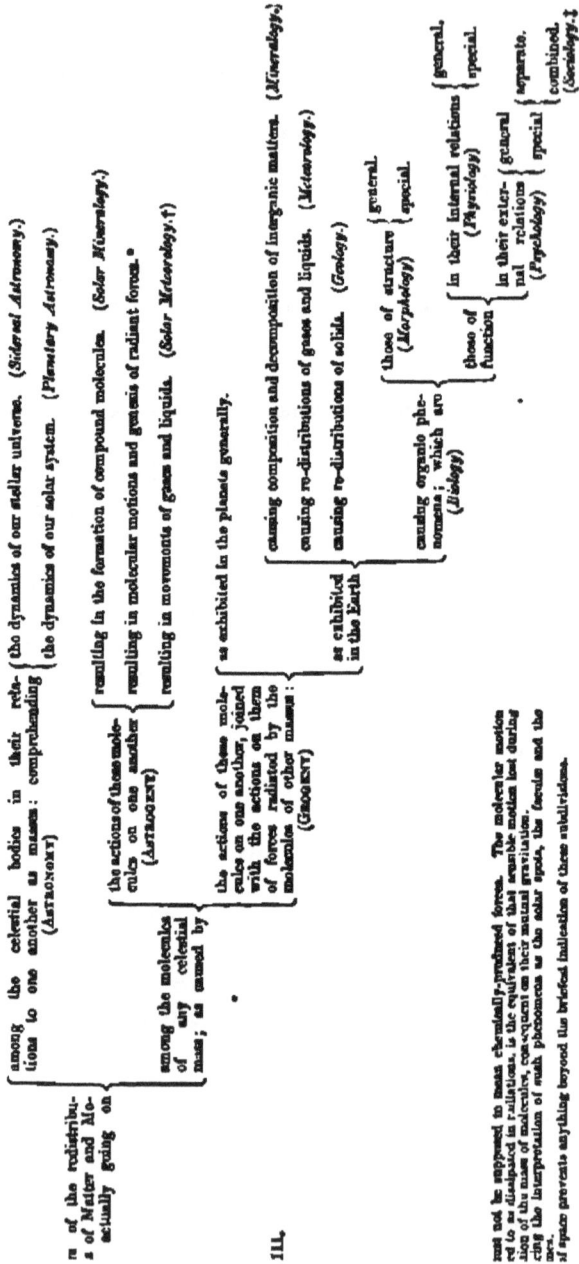

n of the redistribu- s of Matter and Mo- actually going on

- among the celestial bodies in their rela- tions to one another as masses: comprehending (ASTRONOMY)
  - the dynamics of our stellar universe. (*Sidereal Astronomy.*)
  - the dynamics of our solar system. (*Planetary Astronomy.*)

- among the molecules of any celestial mass; as caused by
  - the actions of these mole- cules on one another (ASTROGENY)
    - resulting in the formation of compound molecules. (*Solar Mineralogy.*)
    - resulting in molecular motions and genesis of radiant forces.*
    - resulting in movements of gases and liquids. (*Solar Meteorology.*†)

III.

- the actions of these mole- cules on one another, joined with the actions on them of forces radiated by the molecules of other masses: (GEOGENY)
  - as exhibited in the planets generally.
  - as exhibited in the Earth
    - causing composition and decomposition of inorganic matter. (*Mineralogy.*)
    - causing re-distributions of gases and liquids. (*Meteorology.*)
    - causing re-distributions of solids. (*Geology.*)
    - causing organic phe- nomena; which are (*Biology*)
      - those of structure (*Morphology*)
        - general
        - special
      - those of function
        - in their internal relations (*Physiology*)
          - general
          - special
        - in their exter- nal relations (*Psychology*)
          - general
          - special { separate. combined. (*Sociology.*‡)

must not be supposed to mean chemically-produced forces. The molecular motion red to as dissipated in radiations, is the equivalent of that sensible motion lost during ation of the mass of molecules, consequent on their mutual gravitation.
cing the interpretation of such phenomena as the solar spots, the faculæ and the
ies.
f space prevents saying anything beyond the briefest indication of these subdivisions.

of the components, or than can exist between the
Sciences which deal with one or other order of the
things composed. The three groups of Sciences may
bo briefly defined as—laws of the *forms;* laws of
the *factors;* laws of the *products.* And when thus
defined, it becomes manifest that the groups are
so radically unlike in their natures, that there can
be no transitions between them; and that any
Science belonging to one of the groups must be
quite incongruous with the Sciences belonging to
either of the other groups, if transferred. How
fundamental are the differences between them, will be
further seen on considering their functions. The first,
or abstract group, is *instrumental* with respect to both
the others; and the second, or abstract-concrete group,
is *instrumental* with respect to the third or concrete
group. An endeavour to invert these functions will
at once show how essential is the difference of
character. The second and third groups supply
subject-matter to the first, and the third supplies
subject-matter to the second; but none of the truths
which constitute the third group are of any use as
solvents of the problems presented by the second
group; and none of the truths which the second
group formulates can act as solvents of problems
contained in the first group. Concerning the sub-
divisions of these great groups, little remains to be
added. That each of the groups, being co-extensive
with all phenomena, contains truths that are universal

and others that are not universal, and that these must be classed apart, is obvious. And that the sub-divisions of the non-universal truths, are to be made in something like the manner shown in the tables, is proved by the fact that when the descriptive words are read from the root to the extremity of any branch, they form a definition of the Science constituting that branch. That the minor divisions might be other-wise arranged, and that better definitions of them might be given, is highly probable. They are here set down merely for the purpose of showing how this method of classification works out.

I will only further remark, that the relations of the Sciences as thus represented, are still but imperfectly represented : their relations cannot be truly shown on a plane, but only in space of three dimensions. The three groups cannot rightly be · put in linear order as they have here been. Since the first stands related to the third, not only indirectly through the second, but also directly—it is directly instrumental with respect to the third, and the third supplies it directly with subject-matter. Their relations can thus only be truly shown by a divergence from a common root on different sides, in such a way that each stands in juxta-position to the other two. And only by the like mode of arrangement, can the relations among the sub-divisions of each group be correctly represented.

POSTSCRIPT,

REPLYING TO CRITICISMS.

AMONG objections made to any doctrine, those which come
from avowed supporters of an adverse doctrine must be considered, other things equal, as of less weight than those
which come from men uncommitted to an adverse doctrine,
or but partially committed to it. The element of preposses-
sion, distinctly present in the one case and in the other case
mainly or quite absent, is a well-recognized cause of difference in the values of the judgments: supposing the judgments to be otherwise fairly comparable. Hence, when it is
needful to bring the replies within a restricted space, a fit
course is that of dealing rather with independent criticisms
than with criticisms which are really indirect arguments for
an opposite view, previously espoused.

For this reason I propose here to confine myself substantially, though not absolutely, to the demurrers entered against
the foregoing classification by Prof. Bain, in his recent work
on Logic. Before dealing with the more important of those,
let me clear the ground by disposing of the less important.

Incidentally, while commenting on the view I take respecting the position of Logic, Prof. Bain points out that
this, which is the most abstract of the sciences, owes much
to Psychology, which I place among the Concrete Sciences;
and he alleges an incongruity between this fact and my
statement that the Concrete Sciences are not instrumental

2a

in disclosing the truths of the Abstract Sciences. Subsequently he re-raises this apparent anomaly when saying—

"Nor is it possible to justify the placing of Psychology wholly among Concrete Sciences. It is a highly analytic science, as Mr. Spencer thoroughly knows."

For a full reply, given by implication, I must refer Prof. Bain to § 56 of *The Principles of Psychology*, where I have contended that "while, under its objective aspect, Psychology is to be classed as one of the Concrete Sciences which successively decrease in scope as they increase in speciality; under its subjective aspect, Psychology is a totally unique science, independent of, and antithetically opposed to, all other sciences whatever." A pure idealist will not, I suppose, recognize this distinction; but to every one else it must, I should think, be obvious that the science of subjective existences is the correlative of all the sciences of objective existences; and is as absolutely marked off from them as subject is from object. Objective Psychology, which I class among the Concrete Sciences, is purely synthetic, so long as it is limited, like the other sciences, to objective data; though great aid in the interpretation of these data is derived from the observed correspondence between the phenomena of Objective Psychology as presented in other beings and the phenomena of Subjective Psychology as presented in one's own consciousness. Now it is Subjective Psychology only which is analytic, and which affords aid in the development of Logic. This being explained, the apparent incongruity disappears.

A difficulty raised respecting the manner in which I have expressed the nature of Mathematics, may next be dealt with. Prof. Bain writes:—

"In the first place, objection may be taken to his language, in discussing the extreme Abstract Sciences, when he speaks of the *empty forms* therein considered. To call Space and Time empty forms, must mean that they can be thought of without any concrete embodiment whatsoever; that one can think of Time, as a pure abstraction, without having in one's mind any concrete succession. Now, this doctrine is in the last degree questionable."

I quite agree with Prof. Bain that "this doctrine is in the last degree questionable;" but I do not admit that this doctrine is implied by the definition of Abstract Science which I have given. I speak of Space and Time as they are dealt with by mathematicians, and as it is alone possible for pure Mathematics to deal with them. While Mathematics habitually uses in its points, lines, and surfaces, certain existences, it habitually deals with these as representing points, lines, and surfaces that are ideal; and *its conclusions are true only on condition that it does this.* Points having dimensions, lines having breadths, planes having thicknesses, are negatived by its definitions. Using, though it does, material representatives of extension, linear, superficial, or solid, Geometry deliberately ignores their materiality; and attends only to the truths of relation they present. Holding with Prof. Bain, as I do, that our consciousness of Space is disclosed by our experiences of Matter—arguing, as I have done in *The Principles of Psychology*, that it is a consolidated aggregate of all relations of co-existence that have been severally presented by Matter; I nevertheless contend that it is possible to dissociate these relations from Matter to the extent required for formulating them as abstract truths. I contend, too, that this separation is of the kind habitually made in other cases; as, for instance, when the general laws of motion are formulated (as M. Comte's system, among others, formulates them) in such way as to ignore all properties of the bodies dealt with save their powers of taking up, and retaining, and giving out, quantities of motion; though these powers are inconceivable apart from the attribute of extension, which is intentionally disregarded.

Taking other of Prof. Dain's objections, not in the order in which they stand but in the order in which they may be most conveniently dealt with, I quote as follows:—

"The law of the radiation of light (the inverse square of the distance) is said by Mr. Spencer to be Abstract-Concrete, while the disturbing changes in the medium are not to be mentioned except in a Concrete Science of Optics. We need not remark that such a separate handling is unknown to science."

It is perfectly true that "such a separate handling is unknown to science." But, unfortunately for the objection, it is also perfectly true that no such separate handling is proposed by me, or is implied by my classification. How Prof. Bain can have so missed the meaning of the word "concrete," as I have used it, I do not understand. After pointing out that "no one ever drew the line," between the Abstract-Concrete and the Concrete Sciences, "as I have done it," he alleges an anomaly which exists only supposing that I have drawn it where it is ordinarily drawn. He appears inadvertently to have carried with him M. Comte's conception of Optics as a Concrete Science, and, importing it into my classification, debits me with the incongruity. If he will re-read the definition of the Abstract-Concrete Sciences, or study their sub-divisions as shown in Table II., he will, I think, see that the most special laws of the redistribution of light, equally with its most general laws, are included. And if he will pass to the definition and the tabulation of the Concrete Sciences, he will, I think, see no less clearly that Optics cannot be included among them.

Prof. Bain considers that I am not justified in classing Chemistry as an Abstract-Concrete Science, and excluding from it all consideration of the crude forms of the various substances dealt with; and he enforces his dissent by saying that chemists habitually describe the ores and impure mixtures in which the elements, etc., are naturally found. Undoubtedly chemists do this. But do they therefore intend to include an account of the ores of a substance, *as a part of the science* which formulates its molecular constitution and the constitutions of all the definite compounds it enters into? I shall be very much surprised if I find that they do. Chemists habitually prefix to their works a division treating of Molecular Physics; but they do not therefore claim Molecular Physics as a part of Chemistry. If they similarly prefix to the chemistry of each substance an outline of its mineralogy, I do not think they therefore mean to assert that the last belongs to the first. Chemistry proper, embraces nothing beyond an account of the constitutions and modes of action and combining proportions of substances that are taken as absolutely pure; and its truths no more recognize impure substances than the truths of Geometry recognize crooked lines.

Immediately after, in criticizing the fundamental distinction I have made between Chemistry and Biology, as Abstract-Concrete and Concrete respectively, Prof. Bain says:—

"But the objects of Chemistry and the objects of Biology are equally concrete, so far as they go; the simple bodies of chemistry, and their several compounds, are viewed by the Chemist as concrete wholes, and are described by him, not with reference to one factor, but to all their factors."

Issue is here raised in a form convenient for elucidation of the general question. It is true that, *for purposes of identification*, a chemist gives an account of all the sensible characters of a substance. He sets down its crystalline form, its specific gravity, its power of refracting light, its behaviour as magnetic or diamagnetic. But does he thereby include these phenomena as part of the Science of Chemistry? It seems to me that the relation between the weight of any portion of matter and its bulk, which is ascertained on measuring its specific gravity, is a physical and not a chemical fact. I think, too, that the physicist

will claim, as part of his science, all investigations touching
the refraction of light: be the substance producing this
refraction what it may. And the circumstance that the
chemist may test the magnetic or diamagnetic property
of a body, as a means of ascertaining what it is, or as a
means of helping other chemists to determine whether they
have got before them the same body, will neither be held
by the chemist, nor allowed by the physicist, to imply a
transfer of magnetic phenomena from the domain of the
one to that of the other. In brief, though the chemist, in
his account of an element or a compound, may refer to
certain physical traits associated with its molecular constitution and affinities, he does not by so doing change these
into chemical traits. Whatever chemists may put into
their books, Chemistry, considered as a science, includes
only the phenomena of molecular structures and changes—
of compositions and decompositions.* I contend, then,
that Chemistry does *not* give an account of anything
as a concrete whole, in the same way that Biology gives
an account of an organism as a concrete whole. This
will become even more manifest on observing the character
of the biological account. All the attributes of an organism
are comprehended, from the most general to the most special
—from its conspicuous structural traits to its hidden and faint
ones; from its outer actions that thrust themselves on the
attention, to the minutest sub-divisions of its multitudinous

* Perhaps some will say that such incidental phenomena as those of the heat
and light evolved during chemical changes, are to be included among chemical
phenomena. I think, however, the physicist will hold that all phenomena of
re-distributed molecular motion, no matter how arising, come within the range
of Physics. But whatever difficulty there may be in drawing the line between
Physics and Chemistry (and, as I have incidentally pointed out in *The Principles
of Psychology*, § 65, the two are closely linked by the phenomena of allotropy
and isomerism), applies equally to the Comtean classification, or to any other.
And I may further point out that no obstacle hence arises to the classification I
am defending. Physics and Chemistry being both grouped by me as Abstract-
Concrete Sciences, no difficulty in satisfactorily dividing them in the least affects
the satisfactoriness of the division of the great group to which they both belong,
from the other two great groups.

internal functions; from its character as a germ, through the many changes of size, form, organization, and habit, it goes through until death; from the physical characters of it as a whole, to the physical characters of its microscopic cells, and vessels, and fibres; from the chemical characters of its substance in general to the chemical characters of each tissue and each secretion—all these, with many others. And not only so, but there is comprehended as the ideal goal of the science, the *consensus* of all these phenomena in their co-existences and successions, as constituting a coherent individualized group definitely combined in space and in time. It is this recognition of *individuality* in its subject-matter, that gives its concreteness to Biology, as to every other Concrete Science. As Astronomy deals with bodies that have their several proper names, or (as with the smaller stars) are registered by their positions, and considers each of them as a distinct individual—as Geology, while dimly perceiving in the Moon and nearest planets other groups of geological phenomena (which it would deal with as independent wholes, did not distance forbid), occupies itself with that individualized group presented by the Earth; so Biology treats either of an individual distinguished from all others, or of parts or products belonging to such an individual, or of structural or functional traits common to many such individuals that have been observed, and supposed to be common to others that are like them in most or all of their attributes. Every biological truth connotes a specifically individualized object, or a number of specifically individualized objects of the same kind, or numbers of different kinds that are severally specific. See, then, the contrast. The truths of the Abstract-Concrete Sciences do not imply specific individuality. Neither Molar Physics, nor Molecular Physics, nor Chemistry, concerns itself with this. The laws of motion are expressed without any reference whatever to the sizes or shapes of the moving masses; which may be taken indifferently to be suns or atoms. The relations between contraction and the escape of molecular motion, and between expansion and the absorption of molecular motion, are expressed in their general forms without reference to the kind of matter; and, if the degree of either that occurs in a particular kind of matter is formulated, no note is taken of the quantity of that matter, much less of its individuality. Similarly with Chemistry. When it inquires into the atomic weight, the molecular structure, the atomicity, the combining proportions, etc., of a substance, it is indifferent whether a grain or a ton be thought of—the conception of amount is absolutely irrelevant. And so with more special attributes. Sulphur, considered chemically, is not sulphur under its crystalline form, or under its allotropic viscid form, or as a liquid, or as a gas; but it is sulphur considered apart from those attributes of quantity, and shape, and state, that give individuality.

Prof. Bain objects to the division I have drawn between the Concrete Science of Astronomy and that Abstract-Concrete Science which deals with the mutually-modified motions of hypothetical masses in space, as "not a little arbitrary." He says:—

"We can suppose a science to confine itself *solely* to the 'factors,' or the separated elements, and never, on any occasion, to combine two into a composite third. This position is intelligible, and possibly defensible. For example, in Astronomy, the Law of Persistence of Motion in a straight line might be discussed in pure ideal separation; and so, the Law of Gravity might be discussed in equally pure separation—both under the Abstract-Concrete department of Mechanics. It might then be reserved to a *concrete* department to unite these in the explanation of a projectile or of a planet. Such, however, is not Mr. Spencer's boundary line. He allows Theoretical Mechanics to make this particular combination, and to arrive at the laws of planetary movement, *in the case of a single planet*. What he does not allow is, to proceed to the case of two planets, mutually disturbing one another, or a planet and a satellite, commonly called the 'problem of the Three Bodies.'"

If I held what Prof. Bain supposes me to hold, my position would be an absurd one; but he misapprehends me. The misapprehension results in part from his having here, as before, used the word "concrete" with the Comtean meaning, as though it were my meaning; and in part from the inadequacy of my explanation. I did not in the least mean to imply that the Abstract-Concrete Science of Mechanics, when dealing with the motions of bodies in space, is limited to the interpretation of planetary movement such as it would be did only a single planet exist. It never occurred to me that my words (see p. 19) might be so construed. Abstract-Concrete problems admit, in fact, of being complicated indefinitely, without going in the least beyond the definition. I do not draw the line, as Prof. Bain alleges, between the combination of two factors and the combination of three, or between the combination of any number and any greater number. I draw the line between the science which deals with the theory of the factors, taken singly and in combinations of two, three, four, or more, and the science which, *giving to these factors the values derived from observations of actual objects, uses the theory to explain actual phenomena.*

It is true that, in these departments of science, no radical distinction is consistently recognized between theory and the applications of theory. As Prof. Bain says:—

"Newton, in the First Book of the Principia, took up the problem of the Three Bodies, as applied to the Moon, and worked it to exhaustion. So writers on Theoretical Mechanics continue to include the Three Bodies, Precession, and the Tides."

But, supreme though the authority of Newton may be as a mathematician and astronomer, and weighty as are the names of Laplace and Herschel, who in their works have similarly mingled theorems and the explanations yielded by them, it does not seem to me that these facts go for much; unless it can be shown that these writers intended thus to enunciate the views at which they had arrived respecting the classifi-

cation of the sciences. Such a union as that presented in
their works, adopted merely for the sake of convenience, is,
in fact, the indication of incomplete development; and has
been parallelod in simpler sciences which have afterwards
outgrown it. Two conclusive illustrations are at hand. The
name Geometry, utterly inapplicable by its meaning to the
science as it now exists, was applicable in that first stage
when its few truths were taught in preparation for land-
measuring and the setting-out of buildings; but, at a comparatively early date, those comparatively simple truths
became separated from their applications, and were embodied
by the Greek geometers into systems of theory.* A like puri-
fication is now taking place in another division of the science.
In the *Géométrie Descriptive* of Monge, theorems were mixed
with their applications to projection and plan-drawing. But,
since his time, the science and the art have been segregating;
and Descriptive Geometry, or, as it may be better termed,
the Geometry of Position, is now recognized by mathematicians as a far-reaching system of truths, parts of which are
already embodied in books that make no reference to derived
methods available by the architect or the engineer. To meet
a counter-illustration that will be cited, I may remark that
though, in works on Algebra intended for beginners, the
theories of quantitative relations, as treated algebraically,
are accompanied by groups of problems to be solved, the
subject-matters of these problems are not thereby made
parts of the Science of Algebra. To say that they are, is
to say that Algebra includes the conceptions of distances
and relative speeds and times, or of weights and bulks
and specific gravities, or of areas ploughed and days and
wages; since these, and endless others, may be the terms of

---

* It may be said that the mingling of problems and theorems in Euclid is not
quite consistent with this statement; and it is true that we have, in this mingling,
a trace of the earlier form of the science. But it is to be remarked that these
problems are all purely abstract, and, further, that each of them admits of being
expressed as a theorem.

its equations. And just in the same way that these concrete problems, solved by its aid, cannot by any possibility be incorporated with the Abstract Science of Algebra; so I contend that the concrete problems of Astronomy, cannot by any possibility be incorporated with that division of Abstract-Concrete Science which develops the theory of the inter-actions of free bodies that attract one another.

On this point I find myself at issue, not only with Prof. Bain, but also with Mr. Mill, who contends that :—

"There *is* an abstract science of astronomy, namely, the theory of gravitation, which would equally agree with and explain the facts of a totally different solar system from the one of which our earth forms a part. The actual facts of our own system, the dimensions, distances, velocities, temperatures, physical constitution, etc., of the sun, earth, and planets, are properly the subject of a concrete science, similar to natural history; but the concrete is more inseparably united to the abstract science than in any other case, since the few celestial facts really accessible to us are nearly all required for discovering and proving the law of gravitation as an universal property of bodies, and have therefore an indispensable place in the abstract science as its fundamental data."—*Auguste Comte and Positivism*, p. 43.

In this explanation, Mr. Mill recognizes the fundamental distinction between the Concrete Science of Astronomy, dealing with the bodies actually distributed in space, and a science dealing with hypothetical bodies hypothetically distributed in space. Nevertheless, he regards those sciences as not separable; because the second derives from the first the data whence the law of inter-action is derived. But the truth of this premiss, and the legitimacy of this inference, may alike be questioned. The discovery of the law of inter-action was not due primarily, but only secondarily, to observation of the heavenly bodies. The conception of an inter-acting force that varies inversely as the square of the distance, is an *à priori* conception rationally deducible from mechanical and geometrical considerations. Though unlike in derivation to the many empirical hypotheses of Kepler respecting planetary orbits and planetary motions, yet it was like the successful among these in its relation to astronomical phenomena: it was one of many possible hypotheses, which admitted of having their consequences worked out and tested; and one which, on having its implications compared with the results of observation, was found to explain them. In short, the theory of gravitation grew out of experiences of terrestrial phenomena; but the verification of it was reached through experiences of celestial phenomena. Passing now from premiss to inference, I do not see that, even were the alleged parentage substantiated, it would necessitate the supposed inseparability; any more than the descent of Geometry from land-measuring necessitates a persistent union of the two. In the case of Algebra, as above indicated, the disclosed laws of quantitative relations hold throughout multitudinous orders of phenomena that are extremely heterogeneous; and this makes conspicuous the distinction between the theory and its applications. Here the laws of quantitative relations among masses, distances, velocities, and momenta, being applied mainly (though not exclusively) to the concrete cases presented by Astronomy, the distinction between the theory and its applications is less conspicuous. But, intrinsically, it is as great in the one case as in the other.

How great it is, we shall see on taking an analogy. This is a living man, of whom we may know little more than that he is a visible, tangible person; or of whom we may know enough to form a voluminous biography. Again, this book tells of a fictitious hero, who, like the heroes of old romance, may be an impersonated virtue or vice, or, like a modern hero, one of mixed nature, whose various motives and consequent actions are elaborated into a semblance of reality. But no accuracy and completeness of the picture makes this fictitious personage an actual personage, or brings him any nearer to one. Nor does any meagreness in our knowledge of a real man reduce him any nearer to the imaginary being of a novel. To the last, the division between fiction and biography remains an impassable gulf. So, too, remains the division between the Science dealing with the inter-actions of hypothetical bodies in space, and the Science dealing with the inter-actions of existing bodies in space. We may elaborate the first to any degree whatever by the introduction of three, four, or any greater number of factors under any number of assumed conditions, until we symbolize a solar system; but to the last an account of our symbolic solar system is as far from an account of the actual solar system as fiction is from biography.

Even more obvious, if it be possible, does the radical character of this distinction become, on observing that from the simplest proposition of General Mechanics we may pass to the most complex proposition of Celestial Mechanics, without a break. We take a body moving at a uniform velocity, and commence with the proposition that it will continue so to move for ever. Next, we state the law of its accelerated motion in the same line, when subject to a uniform force. We further complicate the proposition by supposing the force to increase in consequence of approach towards an attracting body; and we may formulate a series of laws of acceleration, resulting from so many assumed laws of increasing attraction (of which the law of gravitation is one). Another factor may now be added by supposing the body to have motion in a direction other than that of the attracting body; and we may determine, according to the ratios of the supposed forces, whether its course will be hyperbolic, parabolic, elliptical, or circular—we may begin with this hypothetical additional force as infinitesimal, and formulate the varying results as it is little by little increased. The problem is complicated a degree more by taking into account the effects of a third force, acting in some other direction; and beginning with an infinitesimal amount of this force we may reach any amount. Similarly, by introducing factor after factor, each at first insensible in proportion to the rest, we arrive, through an infinity of gradations, at a combination of any complexity.

Thus, then, the Science which deals with the inter-action of hypothetical bodies in space, is *absolutely continuous* with General Mechanics. We have already seen that it is *absolutely discontinuous* with that account of the heavenly bodies which has been called Astronomy from the beginning. When these facts are recognized, it seems to me that there cannot remain a doubt respecting its true place in a classification of the Sciences.

Passing over minor criticisms, either as met by implication or as demanding space that cannot be here afforded, let me say something by way of enforcing the general argument. I will re-state the case in two ways: the first of them adapted only to those who accept the general doctrine of Evolution.

We set out with concentrating nebulous matter. Tracing the re-distributions of this as the rotating contracting spheroid leaves behind successive annuli, and as those severally breaking up eventually form secondary rotating spheroids, we come at length to planets in their early stages. Thus far we consider the phenomena dealt with purely astronomical; and so long as our Earth, regarded as one of these spheroids, was made up of gaseous and molten matters only, it presented no definite data for any more complex Concrete Science. In the lapse of cosmical time a solid film forms, which, in the course of millions of years, thickens, and, in the course of further millions of years, becomes cool enough to permit the precipitation, first of various other gaseous compounds, and finally of water. Presently, the varying exposure of different parts of the spheroid to the Sun's rays, begins to produce appreciable effects; until at length there have arisen meteorological actions, and consequent geological actions, such as those we now know: determined partly by the Sun's heat, partly by the still-retained internal heat of the Earth, and partly by the action of the Moon on the ocean? How have we reached these geological phenomena? When did the astronomical changes end and the geological begin? It needs but to ask this question to see that there is no real division between the two. Putting pre-conceptions aside, we find nothing more than a group of phenomena continually complicating under the influence of the same original factors; and we see that our conventional division is defensible only on grounds of convenience. Let us advance a stage. As the Earth's surface continues to cool, passing through all degrees of temperature by infinitesimal gradations, the formation of more and more complex inorganic compounds becomes possible; later its surface sinks to that heat at which the less complex compounds of the kinds called organic can exist; and finally the formation of the more complex organic compounds becomes possible. Chemists now show us that these compounds may be built up synthetically in the laboratory—each stage in ascending complexity making possible the next higher stage. Hence it is inferable that, in the myriads of laboratories, endlessly diversified in their materials and conditions, which the Earth's surface furnished during the myriads of years occupied in passing through these stages of temperature, such successive syntheses were effected; and that the highly complex unstable substance out of which all organisms are composed, was eventually formed in microscopic portions: from which, by continuous integrations and differentiations, the evolution of all organisms has proceeded. Where then shall we draw the line between Geology and Biology? The synthesis of this most complex compound, is but a continuation of the syntheses by which all simpler compounds were formed.

The same primary factors have been co-operating with those secondary factors, meteorologic and geologic, previously derived from them. Nowhere do we find a break in the ever-complicating series; for there is a manifest connexion between those movements which various complex compounds undergo during their isomeric transformations, and those changes of form undergone by the protoplasm which we distinguish as living. Strongly contrasted as they eventually become, biological phenomena are at their root inseparable from geological phenomena—inseparable from the aggregate of transformations continually wrought in the matters forming the Earth's surface by the physical forces to which they are exposed. Further stages I need not particularize. The gradual development out of the biological group of phenomena, of the more specialized group we class as psychological, needs no illustration. And when we come to the highest psychological phenomena, it is clear that since aggregations of human beings may be traced upwards from single wandering families to tribes and nations of all sizes and complexities, we pass insensibly from the phenomena of individual human action to those of corporate human action. To resume, then, is it not manifest that in the group of sciences—Astronomy, Geology, Biology, Psychology, Sociology, we have a natural group that admits neither of disruption nor change of order? Here there is both a genetic dependence, and a dependence of interpretations. The phenomena have arisen in this succession in cosmical time; and complete scientific interpretation of each group depends on scientific interpretation of the preceding groups. No other science can be thrust in anywhere without destroying the continuity. To insert Physics between Astronomy and Geology, would be to make a break in the history of a continuous series of changes; and a like break would be produced by inserting Chemistry between Geology and Biology. It is true that Physics and Chemistry are

needful as interpreters of these successive assemblages of
facts; but it does not therefore follow that they are themselves to be placed among these assemblages.

Concrete Science, made up of these five concrete sub-
sciences, being thus coherent within itself, and separated
from all other science, there comes the question—Is all other
science similarly coherent within itself? or is it traversed by
some second division that is equally decided? It is thus
traversed. A statical or dynamical theorem, however simple,
has always for its subject-matter something that is conceived
as extended, and as displaying force or forces—as being a
seat of resistance, or of tension, or of both, and as capable
of possessing more or less of *vis viva*. If we examine the
simplest proposition of Statics, we see that the conception of
Force must be joined with the conception of Space, before
the proposition can be framed in thought; and if we similarly examine the simplest proposition in Dynamics, we see
that Force, Space, and Time, are its essential elements. The
amounts of the terms are indifferent; and, by reduction of
its terms beyond the limits of perception, they are applied to
molecules: Molar Mechanics and Molecular Mechanics are
continuous. From questions concerning the relative motions
of two or more molecules, Molecular Mechanics passes to
changes of aggregation among many molecules, to changes
in the amounts and kinds of the motions possessed by them
as members of an aggregate, and to changes of the motions
transferred through aggregates of them (as those constituting
light). Daily extending its range of interpretations, it is
coming to deal even with the components of each compound
molecule on the same principles. And the unions and dis-
unions of such more or less compound molecules, which
constitute the phenomena of Chemistry, are also being conceived as resultant phenomena of essentially kindred natures
—the affinities of molecules for one another, and their re-
actions in relation to light, heat, and other modes of force,

being regarded as consequent on the combinations of the
various mechanically-determined motions of their various
components. Without at all out-running, however, this progress in the mechanical interpretation of molecular phenomena, it suffices to point out that the indispensable
elements in any chemical conception are units occupying
places in space, and exerting forces on one another. This,
then, is the common character of all these sciences which
we at present group under the names of Mechanics,
Physics, Chemistry. Leaving undiscussed the question
whether it is possible to conceive of force apart from extended somethings exerting it, we may assert, as beyond
dispute, that if the conception of force be expelled, no
science of Mechanics, Physics, or Chemistry remains. Made
coherent, as these sciences are, by this bond of union, it is
impossible to thrust among them any other science without
breaking their continuity. We cannot place Logic between
Molar Mechanics and Molecular Mechanics. We cannot place
Mathematics between the group of propositions concerning
the behaviour of homogeneous molecules to one another, and
the group of propositions concerning the behaviour of heterogeneous molecules to one another (which we call Chemistry).
Clearly these two sciences lie outside the coherent whole we
have contemplated : separated from it in some radical way.

By what are they radically separated ? By the absence of
the conception of force. However true it may be that so
long as Logic and Mathematics have any terms at all, these
must be capable of affecting consciousness, and, by implication, of exerting force ; yet it is the distinctive trait of
these sciences that not only do their propositions make no
reference to such force, but, as far as possible, they deliberately ignore it. Instead of being, as in all the other
sciences, an element that is not only recognized but vital ; in
Mathematics and Logic, force is an element that is not only
not vital, but is studiously not recognized. The terms in

which Logic expresses its propositions, are symbols that do not profess to represent things, properties, or powers, of one kind more than another; and may equally well stand for the attributes belonging to members of some connected series of ideal curves which have never been drawn, as for so many real objects. And the theorems of Geometry, so far from contemplating perceptible lines and surfaces as elements in the truths enunciated, consider these truths as becoming absolute only when such lines and surfaces become ideal—only when the conception of something exercising force is extruded.

Let me now make a second re-statement, not implying acceptance of the doctrine of Evolution, but exhibiting with a clearness almost if not quite as great, these fundamental distinctions.

The concrete sciences, taken together or separately, contemplate as their subject-matters, *aggregates*—either the entire aggregate of sensible existences, or some secondary aggregate separable from this entire aggregate, or some tertiary aggregate separable from this, and so on. Sidereal Astronomy occupies itself with the totality of visible masses distributed through space; which it deals with as made up of identifiable individuals occupying specified places, and severally standing towards one another, towards sub-groups, and towards the entire group, in defined ways. Planetary Astronomy, cutting out of this all-including aggregate that relatively minute part constituting the Solar System, deals with this as a whole—observes, measures, and calculates the sizes, shapes, distances, motions, of its primary, secondary, and tertiary members; and, taking for its larger inquiries the mutual actions of all these members as parts of a co-ordinated assemblage, takes for its smaller inquiries the actions of each member considered as an individual, having a set of intrinsic activities that are modified by a set of extrinsic activities. Restricting itself to one of these aggregates, which admits of close examination, Geology (using this word in its comprehensive meaning) gives an account of terrestrial actions and terrestrial structures, past and present; and, taking for its narrower problems local formations and the agencies to which they are due, takes for its larger problems the aerial transformations undergone by the entire Earth. The geologist being occupied with this cosmically small, but otherwise vast, aggregate, the biologist occupies himself with small aggregates formed out of parts of the Earth's superficial substance, and treats each of these as a coordinated whole in its structures and functions; or, when he treats of any particular organ, considers this as a whole made up of parts held in a sub-coordination that refers to the coordination of the entire organism. To the psychologist he leaves those specialized aggregates of functions which adjust the actions of organisms to the complex activities surrounding them: doing this, not simply because they are a stage higher in speciality, but because they are the counterparts of those aggregated states of consciousness dealt with by the science of Subjective Psychology, which stands entirely apart from all other sciences. Finally, the sociologist considers each tribe and nation as an aggregate presenting multitudinous phenomena, simultaneous and successive, that are held together as parts of one combination. Thus, in every case, a concrete science deals with a real aggregate (or a plurality of such aggregates); and it includes as its subject-matter whatever is to be known of this aggregate in respect of its size, shape, motions, density, texture, general arrangement of parts, minute structure, chemical composition, temperature, etc., together with all the multitudinous changes, material and dynamical, gone through by it from the time it begins to exist as an aggregate to the time it ceases to exist as an aggregate.

No abstract-concrete science makes the remotest attempt

to do anything of this sort. Taken together, the abstract-concrete sciences give an account of the various kinds of *properties* which aggregates display; and each abstract-concrete science concerns itself with a certain order of these properties. By this, the properties common to all aggregates are studied and formulated; by that, the properties of aggregates having special forms, special states of aggregation, etc.; and by others, the properties of particular components of aggregates when dissociated from other components. But by all these sciences the aggregate, considered as an individual object, is tacitly ignored; and a property, or a connected set of properties, exclusively occupies attention. It matters not to Mechanics whether the moving mass it considers is a planet or a molecule, a dead stick thrown into the river or the living dog that leaps after it: in any case the curve described by the moving mass conforms to the same laws. Similarly when the physicist takes for his subject the relation between the changing bulk of matter and the changing quantity of molecular motion it contains. Dealing with the subject generally, he leaves out of consideration the kind of matter; and dealing with the subject specially in relation to this or that kind of matter, he ignores the attributes of size and form: save in the still more special cases where the effect on form is considered, and even then size is ignored. So, too, is it with the chemist. A substance he is investigating, never thought of by him as distinguished in extension or amount, is not even required to be perceptible. A portion of carbon on which he is experimenting, may or may not have been visible under its forms of diamond or graphite or charcoal—this is indifferent. He traces it through various disguises and various combinations—now as united with oxygen to form an invisible gas; now as hidden with other elements in such more complex compounds as ether, and sugar, and oil. By sulphuric acid or other agent he precipitates it from these as a coherent cinder, or as a diffused impalpable powder; and again, by applying heat, forces it to disclose itself as an element of animal tissue. Evidently, while thus ascertaining the affinities and atomic equivalence of carbon, the chemist has nothing to do with any aggregate. He deals with carbon in the abstract, as something considered apart from quantity, form, appearance, or temporary state of combination; and conceives it as the possessor of powers or properties, whence the special phenomena he describes result: the ascertaining of all those powers or properties being his sole aim.

Finally, the Abstract Sciences ignore alike aggregates and the powers which aggregates or their components possess; and occupy themselves with *relations*—either with the relations among aggregates, or among their parts, or the relations among aggregates and properties, or the relations among properties, or the relations among relations. The same logical formula applies equally well, whether its terms are men and their deaths, crystals and their planes of cleavage, or letters and their sounds. And how entirely Mathematics concerns itself with relations, we see on remembering that it has just the same expression for the characters of an infinitesimal triangle, as for those of the triangle which has Sirius for its apex and the diameter of the Earth's orbit for its base.

I cannot see how these definitions of these groups of sciences can be questioned. It is undeniable that every Concrete Science gives an account of an aggregate or of aggregates, inorganic, organic, or super-organic (a society); and that, not concerning itself with properties of this or that order, it concerns itself with the co-ordination of the assembled properties of all orders. It seems to me no less certain that an Abstract-Concrete Science gives an account of some order of properties, general or special; not caring about the other traits of an aggregate displaying them, and not recognizing aggregates at all further than is implied by discussion of the particular order of properties. And I think it is equally clear that an Abstract Science, freeing its propositions, so far as the nature of thought permits, from aggregates and properties, occupies itself with the relations of co-existence and sequence, as disentangled from all particular forms of being and action. If then these three groups of sciences are, respectively, accounts of *aggregates*, accounts of *properties*, accounts of *relations*, it is manifest that the divisions between them are not simply perfectly clear, but that the chasms between them are absolute.

Here, perhaps more clearly than before, will be seen the untenability of the classification made by M. Comte. Already (p. 11), after setting forth in a general way these fundamental distinctions, I have pointed out the incongruities that arise when the sciences, conceived as Abstract, Abstract-Concrete, and Concrete, are arranged in the order proposed by him. Such incongruities become still more conspicuous if for these general names of the groups we substitute the definitions given above. The series will then stand thus:—

MATHEMATICS ....An account of *relations*
  (including, under Mechanics, an account of *properties*).
ASTRONOMY ....An account of *aggregates.*
PHYSICS ....An account of *properties.*
CHEMISTRY ....An account of *properties.*
BIOLOGY ....An account of *aggregates.*
SOCIOLOGY ....An account of *aggregates.*

That those who espouse opposite views see clearly the defects in the propositions of their opponents and not those in their own, is a trite remark that holds in philosophical discussions as in all others: the parable of the mote and

the beam applies as well to men's appreciations of one
another's opinions as to their appreciations of one another's
natures. Possibly to my positivist friends I exemplify this
truth,—just as they exemplify it to me. Those uncommitted to either view must decide where the mote exists and
where the beam. Meanwhile it is clear that one or other
of the two views is essentially erroneous; and that no quali-
fications can bring them into harmony. Either the sciences
admit of no such grouping as that which I have described,
or they admit of no such serial order as that given by
M. Comte.

LONDON,
February, 1871.

REASONS FOR DISSENTING

FROM THE

PHILOSOPHY OF M. COMTE.

WHILE the preceding pages were passing through the press, there appeared in the *Revue des Deux Mondes* for February 15th, an article on a late work of mine—*First Principles*. To M. Auguste Laugel, the writer of this article, I am much indebted for the careful exposition he has made of some of the leading views set forth in that work; and for the catholic and sympathetic spirit in which he has dealt with them. In one respect, however, M. Laugel conveys to his readers an erroneous impression—an impression doubtless derived from what appears to him adequate evidence, and doubtless expressed in perfect sincerity. M. Laugel describes me as being, in part, a follower of M. Comte. After describing the influence of M. Comte as traceable in the works of some other English writers, naming especially Mr. Mill and Mr. Buckle, he goes on to say that this influence, though not avowed, is easily recognizable in the work he is about to make known; and in several places throughout his review, there are remarks having the same implication. I greatly regret having to take exception to anything said by a critic so candid and so able. But the *Revue des Deux Mondes* circulates widely in England, as well as elsewhere; and finding that there exists in some minds, both here and in America, an impression similar to that entertained by M. Laugel— an impression likely to be confirmed by his statement—it appears to me needful to meet it.

Two causes of quite different kinds, have conspired to diffuse the erroneous belief that M. Comte is an accepted exponent of scientific opinion. His bitterest foes and his closest friends, have unconsciously joined in propagating it. On the one hand, M. Comte having designated by the term "Positive Philosophy" all that definitely-established knowledge which men of science have been gradually organizing into a coherent body of doctrine; and having habitually placed this in opposition to the incoherent body of doctrine defended by theologians; it has become the habit of the theological party to think of the antagonist scientific party, under the title of "positivists." And thus, from the habit of calling them "positivists," there has grown up the assumption that they call themselves "positivists," and that they are the disciples of M. Comte. On the other hand, those who have accepted M. Comte's system, and believe it to be the philosophy of the future, have naturally been prone to see everywhere the signs of its progress; and wherever they have found opinions in harmony with it, have ascribed these opinions to the influence of its originator. It is always the tendency of discipleship to magnify the effects of the master's teachings; and to credit the master with all the doctrines he teaches. In the minds of his followers, M. Comte's name is associated with scientific thinking, which, in many cases, they first understood from his exposition of it. Influenced as they inevitably are by this association of ideas, they are reminded of M. Comte wherever they meet with thinking which corresponds, in some marked way, to M. Comte's description of scientific thinking; and hence are apt to imagine him as introducing into other minds, the conceptions which he introduced into their minds. Such impressions are, however, in most cases quite unwarranted. That M. Comte has given a general exposition of the doctrine and method elaborated by Science, is true. But it is not true that the holders of this doctrine and followers of this method, are disciples of M. Comte. Neither their modes of inquiry nor their views concerning human knowledge in its nature and limits, are appreciably different from what they were before. If they are "positivists," it is in the sense that all men of science have been more or less consistently "positivists;" and the applicability of M. Comte's title to them, no more makes them his disciples, than does its applicability to men of science who lived and died before M. Comte wrote, make these his disciples. M. Comte himself by no means claims that which some of his adherents are apt, by implication, to claim for him. He says:—"Il y a, sans doute, beaucoup d'analogie entre ma *philosophie positive* et ce que les savans anglais entendent, depuis Newton surtout, par *philosophie naturelle*;" (see *Avertissement*) and further on he indicates the "grand mouvement imprimé à l'esprit humain, il y a deux siècles, par l'action combinée des préceptes de Bacon, des conceptions de Descartes, et des découvertes de Galileé, comme le moment où l'esprit de la philosophie positive a commencé à se prononcer dans le monde." That is to say, the general mode of thought and way of interpreting phenomena, which M. Comte calls "Positive Philosophy," he recognizes as having been growing for two centuries; as having reached, when he wrote, a marked development; and as being the heritage of all men of science.

That which M. Comte proposed to do, was to give scientific thought and method a more definite embodiment and organization; and to apply it to the interpretation of classes of phenomena not previously dealt with in a scientific manner The conception was a great one; and the endeavour to work it out was worthy of sympathy and applause. Some such conception was entertained by Bacon. He, too, aimed at the organization of the sciences; he, too, held that "Physics is the mother of all the sciences;" he, too, held that the sciences can be advanced only by combining them,

and saw the nature of the required combination; he, too,
held that moral and civil philosophy could not flourish when
separated from their róots in natural philosophy ; and thus
he, too, had some idea of a social science growing out of
physical science. But the state of knowledge in his day pre-
vented any advance beyond the general conception : indeed,
it was marvellous that he should have advanced so far. Instead of a vague, undefined conception, M. Comte has presented the world with a defined and highly-elaborated
conception. In working out this conception he has shown
remarkable breadth of view, great originality, immense fer-
tility of thought, unusual powers of generalization. Considered apart from the question of its truth, his system of
Positive Philosophy is a vast achievement. But after according to M. Comte high admiration for his conception, for
his effort to realize it, and for the faculty he has shown in
the effort to realize it, there remains the inquiry—Has he
succeeded ? A thinker who re-organizes the scientific method
and knowledge of his age, and whoso re-organization is
accepted by his successors, may rightly be said to have such
successors for his disciples. But successors who accept this
method and knowledge of his age, minus his re-organization,
are certainly not his disciples. How then stands the case
with M. Comte ? There are some few who receive his
doctrines with but little reservation ; and these are his disciples truly so called. There are others who regard with
approval certain of his leading doctrines, but not the rest :
these we may distinguish as partial adherents. There
are others who reject all his distinctive doctrines ; and these
must be classed as his antagonists. The members of this
class stand substantially in the same position as they would
have done had he not written. Declining his re-organization of scientific doctrine, they possess this scientific
doctrine in its pre-existing state, as the common heritage
bequeathed by the past to the present ; and their adhesion to

this scientific doctrine in no sense implicates them with M. Comte. In this class stand the great body of men of science. And in this class I stand myself

Coming thus to the personal part of the question, let me first specify those great general principles on which M. Comte is at one with preceding thinkers: and on which I am at one with M. Comte.

All knowledge is from experience, holds M. Comte; and this I also hold—hold it, indeed, in a wider sense than M. Comte: since, not only do I believe that all the ideas acquired by individuals, and consequently all the ideas transmitted by past generations, are thus derived; but I also contend that the very faculties by which they are acquired, are the products of accumulated and organized experiences received by ancestral races of beings (see *Principles of Psychology*). But the doctrine that all knowledge is from experience, is not originated by M. Comte; nor is it claimed by him. He himself says—"Tous les bons esprits répètent, depuis Bacon, qu'il n'y a de connaissances réelle que celles qui reposent sur des faites observés." And the elaboration and definite establishment of this doctrine, has been the special characteristic of the English school of Psychology. Nor am I aware that M. Comte, accepting this doctrine, has done anything to make it more certain, or give it greater definiteness. Indeed it was impossible for him to do so; since he repudiates that part of mental science by which alone this doctrine can be proved.

It is a further belief of M. Comte, that all knowledge is phenomenal or relative; and in this belief I entirely agree. But no one alleges that the relativity of all knowledge was first enunciated by M. Comte. Among others who have more or less consistently held this truth, Sir William Hamilton enumerates, Protagoras, Aristotle, St. Augustin, Boethius, Averroes, Albertus Magnus, Gerson, Leo Hebræus, Melancthon, Scaliger, Francis Piccolomini, Giordano Bruno, Campanella, Bacon, Spinoza, Newton, Kant. And Sir William Hamilton, in his "Philosophy of the Unconditioned," first published in 1829, has given a scientific demonstration of this belief. Receiving it in common with other thinkers, from preceding thinkers, M. Comte has not, to my knowledge, advanced this belief. Nor indeed could he advance it, for the reason already given—he denies the possibility of that analysis of thought which discloses the relativity of all cognition.

M. Comte reprobates the interpretation of different classes of phenomena by assigning metaphysical entities as their causes; and I coincide in the opinion that the assumption of such separate entities, though convenient, if not indeed necessary, for purposes of thought, is, scientifically considered, illegitimate. This opinion is, in fact, a corollary from the last; and must stand or fall with it. But like the last it has been held with more or less consistency for generations. M. Comte himself quotes Newton's favorite saying —"O! Physics, beware of Metaphysics!" Neither to this doctrine, any more than to the preceding doctrines, has M. Comte given a firmer basis. He has simply re-asserted it; and it was out of the question for him to do more. In this case, as in the others, his denial of subjective psychology debarred him from proving that these metaphysical entities are mere symbolic conceptions which do not admit of verification.

Lastly, M. Comte believes in invariable natural laws— absolute uniformities of relation among phenomena. But very many before him have believed in them too. Long familiar even beyond the bounds of the scientific world, the proposition that there is an unchanging order in things, has, within the scientific world, held, for generations, the position of an established postulate : by some men of science recognized only as holding of inorganic phenomena; but recognized by other men of science, as universal. And M. Comte, accepting this doctrine from the past, has left it substantially as it was. Though he has asserted new uniformities, I do not think scientific men will admit that he has so *demonstrated* them, as to make the induction more certain; nor has he deductively established the doctrine, by showing that uniformity of relation is a necessary corollary from the persistence of force, as may readily be shown.

These, then, are the pre-established general truths with which M. Comte sets out—truths which cannot be regarded as distinctive of his philosophy. "But why," it will perhaps be asked, "is it needful to point out this; seeing that no instructed reader supposes these truths to be peculiar to M. Comte?" I reply that though no disciple of M. Comte would deliberately claim them for him; and though no theological antagonist at all familiar with science and philophy, supposes M. Comte to be the first propounder of them; yet there is so strong a tendency to associate any doctrines with the name of a conspicuous recent exponent of them, that false impressions are produced, even in spite of better knowledge. Of the need for making this reclamation, definite proof is at hand. In the No. of the *Revue des Deux Mondes* named at the commencement, may be found, on p. 936, the words—"Toute religion, comme toute philosophie, a la prétention de donner une explication de l'univers. La philosophie qui s'appelle *positive* se distingue de toutes les philosophies et de toutes les religions en ce qu'elle a renoncé à cette ambition de l'esprit humain;" and the remainder of the paragraph is devoted to explaining the doctrine of the relativity of knowledge. The next paragraph begins— "Tout imbu de ces idées, que nous exposons sans les discuter pour le moment, M. Spencer divise, etc." Now this is one of those collocations of ideas which tends to create, or to strengthen, the erroneous impression I would dissipate. I do not for a moment suppose that M. Laugel intended to say that these ideas which he describes as ideas of the "Positive Philosophy," are peculiarly the ideas of M. Comte. But little as ho probably intended it, his expressions suggest this conception. In the minds of both disciples and antagonists, "the Positivo Philosophy" means the philosophy of M. Comte; and to be imbued with the ideas of "the Positive Philosophy" means to be imbued with the ideas of M. Comte —to have received these ideas from M. Comte. After what has been said above, I need scarcely repeat that the conception thus inadvertently suggested, is a wrong one. M. Comte's brief enunciations of these general truths, gave me no clearer apprehensions of them than I had before. Such clarifications of ideas on these ultimate questions, as I can trace to any particular teacher, I owe to Sir William Hamilton

From the principles which M. Comte held in common with many preceding and contemporary thinkers, let us pass now to the principles that are distinctive of his system. Just as entirely as I agree with M. Comte on those cardinal doctrines which wo jointly inherit; so entirely do I disagree with him on those cardinal doctrines which he propounds, and which determine the organization of his philosophy. The best way of showing this will be to compare, side by side, the—

Propositions held by M. Comte.	*Propositions which I hold.*
"... chacune de nos conceptions principales, chaque branche de nos connaissances, passe successivement par trois états théoriques différens: l'état théologique, ou fictif; l'état métaphysique, ou abstrait; l'état scientifique, ou positif. En d'autres termes, l'esprit humain, par sa nature, emploie successivement dans chacune de ses recherches trois méthodes de philoso-	The progress of our conceptions, and of each branch of knowledge, is from beginning to end intrinsically alike. There are not three methods of philosophizing radically opposed; but one method of philosophizing which remains, in essence, the same. At first, and to the last, the conceived causal agencies of phenomena, have a degree of generality corresponding to the width of the generalizations which experiences have determined; and they change just as gradually as experiences accumulate. The inte-

pher, dont le caractère est essentiellement différent et même radicalement opposé : d'abord la méthode théologique, ensuite la méthode métaphysique, et enfin la méthode positive." p. 3.

gration of causal agencies, originally thought of as multitudinous and local, but finally believed to be one and universal, is a process which involves the passing through all intermediate steps between these extremes; and any appearance of stages can be but superficial. Supposed concrete and individual causal agencies, coalesce in the mind as fast as groups of phenomena are assimilated, or seen to be similarly caused. Along with their coalescence, comes a greater extension of their individualities, and a concomitant loss of distinctness in their individualities. Gradually, by continuance of such coalescences, causal agencies become, in thought, diffused and indefinite. And eventually, without any change in the nature of the process, there is reached the consciousness of a universal causal agency, which cannot be conceived.*

' Le système théologique est parvenu à la plus haute perfection dont il soit susceptible, quand il a substitué l'action providentielle d'un être unique au jeu varié des nombreuses divinités indépendantes qui avaient été imaginées primitivement. De même, le dernier terme du système métaphysique consiste à concevoir, au lieu des différentes entités particulières,

As the progress of thought is one, so is the end one. There are not three possible terminal conceptions; but only a single terminal conception. When the theological idea of the providential action of one being, is developed to its ultimate form, by the absorption of all independent secondary agencies, it becomes the conception of a being immanent in all phenomena; and the reduction of it to this state, implies the fading-away, in thought, of all those anthropomorphic attributes by which the aboriginal

* A clear illustration of this process, is furnished by the recent mental integration of Heat, Light, Electricity, etc., as modes of molecular motion. If we go a step back, we see that the modern conception of Electricity, resulted from the integration in consciousness, of the two forms of it evolved in the galvanic battery and in the electric-machine. And going back to a still earlier stage, we see how the conception of statical electricity, arose by the coalescence in thought, of the previously-separate forces manifested in rubbed amber, in rubbed glass, and in lightning. With such illustrations before him, no one can, I think, doubt that the process has been the same from the beginning.

uno seu.e grande entité générale, la *nature*, envisagée comme la source unique de tous les phénomènes. Pareillement, la perfection du système positif, vers laquelle il tend sans cesse, quoiqu'il soit très-probable qu'il ne doive jamais l'atteindre, serait de pouvoir se représenter tous les divers phénomènes observables comme des cas particuliers d'un seul fait général, tel que celui de la gravitation, par exemple." p. 5.

idea was distinguished. The alleged last term of the metaphysical system —the conception of a single great general entity, *nature*, as the source of all phenomena—is a conception identical with the previous one : the consciousness of a single source which, in coming to be regarded as universal, ceases to be regarded as conceivable, differs in nothing but name from the consciousness of one being, manifested in all phenomena. And similarly, that which is described as the ideal state of science—the power to represent all observable phenomena as particular cases of a single general fact, implies the postulating of some ultimate existence of which this single fact is alleged ; and the postulating of this ultimate existence, involves a state of consciousness indistinguishable from the other two.

... la perfection du système positif, vers laquelle il tend sans cesse, quoiqu'il soit très-probable qu'il ne doive jamais l'atteindre, serait de pouvoir se représenter tous les divers phénomènes observables comme des cas particuliers d'un seul fait général. p. 5 ... ... considérant comme absolument inaccessible, et vide de sens pour nous la recherche de ce qu'on appelle les *causes*, soit premières, soit finales." p. 14.

Though along with the extension of generalizations, and concomitant integration of conceived causal agencies, the conceptions of causal agencies grow more indefinite ; and though as they gradually coalesce into a universal causal agency, they cease to be representable in thought, and are no longer supposed to be comprehensible ; yet the consciousness of *cause* remains as dominant to the last as it was at first ; and can never be got rid of. The consciousness of cause can be abolished only by abolishing consciousness itself.* (*First Principles*, § 26.)

---

* Possibly it will be said that M. Comte himself admits, that what he calls the perfection of the positive system, will probably never be reached ; and that what he condemns is the inquiry into the *natures* of causes and not the general recognition of cause. To the first of these allegations, I reply that, as I understand M. Comte, the obstacle to the perfect realization of the positive philosophy is the impossibility of carrying generalization so far as to reduce all particular facts to

37

"Ce n'est pas aux lec-
teurs de cet ouvrage que je
croirai jamais devoir prou-
ver que les idées gouvernent
et bouleversent le monde,
ou, en d'autres termes, que
tout le mécanisme social
repose finalement sur des
opinions. Ils savent surtout
que la grande crise politique
et morale des sociétés ac-
tuelles tient, en dernière
analyse, à l'anarchie intel-
lectuelle." p. 48.*

Ideas do not govern and overthrow
the world : the world is governed or
overthrown by feelings, to which
ideas serve only as guides. The
social mechanism does not rest finally
upon opinions; but almost wholly upon character. Not intellectual anar-
chy, but moral antagonism, is the
cause of political crises. All social
phenomena are produced by the totality of human emotions and beliefs :
of which the emotions are mainly
pre-determined, while the beliefs are
mainly post-determined. Men's desires are chiefly inherited ; but their
beliefs are chiefly acquired, and depend
on surrounding conditions; and the
most important surrounding conditions depend on the social state which
the prevalent desires have produced.
The social state at any time existing,
is the resultant of all the ambitions,
self-interests, fears, reverences, in-
dignations, sympathies, etc., of ancestral citizens and existing citizens.
The ideas current in this social state,
must, on the average, be congruous
with the feelings of citizens ; and
therefore, on the average, with the
social state these feelings have pro-

cases of one general fact—not the impossibility of excluding the consciousness of
cause. And to the second allegation I reply, that the essential principle of his
philosophy, is an avowed ignoring of cause altogether. For if it is not, *what becomes of his alleged distinction between the perfection of the positive system and the
perfection of the metaphysical system ?* And here let me point out that, by affirm-
ing exactly the opposite to that which M. Comte thus affirms, I am excluded
from the positive school. If his own definition of positivism is to be taken,
then, as I hold that what he defines as positivism is an absolute impossibility,
it is clear that I cannot be what he calls a positivist.

* A friendly critic alleges that M. Comte is not fairly represented by this
quotation, and that he is blamed by his biographer, M. Littré, for his too-great
insistance on feeling as a motor of humanity. If in his "Positive Politics,"
which I presume is here referred to, M. Comte abandons his original position, so
much the better. But I am here dealing with what is known as "the Positive
Philosophy;" and that the passage above quoted does not misrepresent it, is
proved by the fact that this doctrine is re-asserted at the commencement of the
Sociology.

duced. Ideas wholly foreign to this social state cannot be evolved, and if introduced from without, cannot get accepted—or, if accepted, die out when the temporary phase of feeling which caused their acceptance, ends. Hence, though advanced ideas when once established, act upon society and aid its further advance; yet the establishment of such ideas depends on the fitness of the society for receiving them. Practically, the popular character and the social state, determine what ideas shall be current; instead of the current ideas determining the social state and the character. The modification of men's moral natures, caused by the continuous discipline of social life, which adapts them more and more to social relations, is therefore the chief proximate cause of social progress. (*Social Statics*, chap. xxx.)

The order in which the generalizations of science are established, is determined by the frequency and impressiveness with which different classes of relations are repeated in conscious experience; and this depends, partly on *the directness with which personal welfare is affected;* partly on *the conspicuousness of one or both the phenomena between which a relation is to be perceived;* partly on *the absolute frequency with which the relations occur;* partly on their *relative frequency of occurrence;* partly on their *degree of simplicity;* and partly on their *degree of abstractness.* (*First Principles*, 1st ed., § 36; appended to this pamphlet.)

"...Je ne dois pas négliger d'indiquer d'avance, comme une propriété essentielle de l'échelle encyclopédique que je vais proposer, sa conformité générale avec l'ensemble de l'histoire scientifique; en ce sens, que, malgré la simultanéité réelle et continue du développement des différentes sciences, celles qui seront classées comme antérieures seront, en effet, plus anciennes et constamment plus avancées que celles présentées comme postérieures." p. 84. . . . . "Cet ordre est déterminé par le degré de simplicité, ou, ce qui revient au même, par le degré de généralité des phénomènes." p. 87.

39

"En résultat définitif, la mathématique, l'astronomie, la physique, la chimie, la physiologie, et la physique sociale; telle est la formule enclyopédique qui, parmi le très-grand nombre de classifications que comportent les six sciences fondamentales, est seule logiquement conforme à la hiérarchie naturelle et invariable des phénomènes." p. 115.

"On conçoit, on effet, que l'étude rationelle de chaque science fondamentale exigeant la culture préalable de toutes celles qui la précèdent dans notre hiérarchie enclyopédique, n'a pu faire de progrès réels et prendre son véritable caractère, qu' après un grand développement des sciences antérieures relatives à des phénomènes plus généraux, plus abstraits, moins compliqués, et indépendans des autres. C'est donc dans cet ordre que la progression, quoique simultanée, a dû avoir lieu." p. 100.

The sciences as arranged in this succession specified by M. Comte, do not logically conform to the natural and invariable hierarchy of phenomena; and there is no serial order whatever in which they can be placed, which represents either their logical dependence or the dependence of phenomena. (See *Genesis of Science*, and foregoing Essay.)

The historical development of the sciences *has not* taken place in this serial order; nor in any other serial order. There is no "true *filiation* of the sciences." From the beginning, the abstract sciences, the abstract-concrete sciences, and the concrete sciences, have progressed together: the first solving problems which the second and third presented, and growing only by the solution of the problems; and the second similarly growing by joining the first in solving the problems of the third. All along there has been a continuous action and reaction between the three great classes of sciences—an advance from concrete facts to abstract facts, and then an application of such abstract facts to the analysis of new orders of concrete facts. (See *Genesis of Science.*)

Such then are the organizing principles of M. Comte's philosophy. Leaving out of his "*Exposition*" those pre-established general doctrines which are the common property of modern thinkers; these are the general doctrines which remain—these are the doctrines which fundamentally distinguish his system. From every one of them I dissent. To each proposition I oppose either a widely-different pro-

40

position, or a direct negation; and I not only do it now, but have done it from the time when I became acquainted with his writings. This rejection of his cardinal principles should, I think, alone suffice; but there are sundry other views of his, some of them largely characterizing his system, which I equally reject. Let us glance at them.

How organic beings have originated, is an inquiry which M. Comte deprecates as a useless speculation: asserting, as he does, that species are immutable.

This inquiry, I believe, admits of answer, and will be answered. That division of Biology which concerns itself with the origin of species, I hold to be the supreme division, to which all others are subsidiary. For on the verdict of Biology on this matter, must wholly depend our conception of human nature, past, present, and future; our theory of the mind; and our theory of society.

M. Comte contends that of what is commonly known as mental science, all that most important part which consists of the subjective analysis of our ideas, is an impossibility.

I have very emphatically expressed my belief in a subjective science of the mind, by writing a *Principles of Psychology*, one half of which is subjective.

M. Comte's ideal of society is one in which *government* is developed to the greatest extent—in which class-functions are far more under conscious public regulation than now—in which hierarchical organization with unquestioned authority shall guide everything—in which the individual life shall be subordinated in the greatest degree to the social life.

That form of society towards which we are progressing, I hold to be one in which *government* will be reduced to the smallest amount possible, and *freedom* increased to the greatest amount possible — one in which human nature will have become so moulded by social discipline into fitness for the social state, that it will need little external restraint, but will be self-restrained—one in which the citizen will tolerate no interference with his freedom, save that which maintains the equal freedom of others —one in which the spontaneous co-operation which has developed our industrial system, and is now developing it with increasing rapidity, will produce agencies for the discharge of nearly all social functions, and will leave to the primary governmental agency nothing beyond the function of maintaining those conditions to free action, which make such spontaneous co-operation possible—one in which individual life will thus be pushed to the greatest extent consistent with social life; and in which social life will have no other end than to maintain the completest sphere for individual life.

M. Comte, not including in his philosophy the consciousness of a cause manifested to us in all phenomena, and yet holding that there must be a religion, which must have an object, takes for his object —Humanity. "This Collective Life (of Society), is in Comte's system the *Être Suprême;* the only one we can *know,* therefore the only one we can worship."

I conceive, on the other hand, that the object of religious sentiment will ever continue to be, that which it has ever been—the unknown source of things. While the *forms* under which men are conscious of the unknown source of things, may fade away, the *substance* of the consciousness is permanent. Beginning with causal agents conceived as imperfectly known; progressing to causal agents conceived as less known and less knowable; and coming at last to a universal causal agent posited as not to be known at all; the religious sentiment must ever continue to occupy itself with this universal causal agent. Having in the course of evolution, come to have for its object of contemplation, the Infinite Unknowable, the religious sentiment can never again (unless by retrogression) take a Finite Knowable, like Humanity, for its object of contemplation.

Here, then, are sundry other points, all of them important, and the last two supremely important, on which I am diametrically opposed to M. Comte; and did space permit, I could add many others. Radically differing from him as I thus do, in everything distinctive of his philosophy; and having invariably expressed my dissent, publicly and privately, from the time I became acquainted with his writings; it may be imagined that I have been not a little startled to find myself classed as one of the same school. That those who have read *First Principles* only, may have been betrayed into this error in the way above shown, by the ambiguous use of the phrase "Positive Philosophy," I can understand. But that any who are acquainted with my previous writings, should suppose I have any general sympathy with M. Comte, save that implied by preferring proved facts to superstitions, astonishes me.

It is true that, disagreeing with M. Comte, though I do, in all those fundamental views that are peculiar to him, I agree with him in sundry minor views. The doctrine that the education of the individual should accord in mode and arrangement with the education of mankind, considered historically, I have cited from him; and have endeavoured to enforce it. I entirely concur in his opinion that there requires a new order of scientific men, whose function shall be that of co-ordinating the results arrived at by the rest. To him I believe I am indebted for the conception of a social *consensus*; and when the time comes for dealing with this conception, I shall state my indebtedness. And I also adopt his word, Sociology. There are, I believe, in the part of his writings which I have read, various incidental thoughts of great depth and value; and I doubt not that were I to read more of his writings, I should find many others.* It is very probable, too, that I have said (as I am told I have) some things which M. Comte had already said. It would be difficult, I believe, to find any two men who had no opinions in common. And it would be extremely strange if two men,

* M. Comte's "Exposition" I read in the original in 1852; and in two or three other places have referred to the original to get his exact words. The Inorganic Physics, and the first chapter of the Biology, I read in Miss Martineau's condensed translation, when it appeared. The rest of M. Comte's views I know only through Mr. Lewes's outline, and through incidental references.

starting from the same general doctrines established by
modern science, should traverse some of the same fields of
inquiry, without their lines of thought having any points
of intersection. But none of these minor agreements can be
of much weight in comparison with the fundamental dis-
agreements above specified. Leaving out of view that general
community which we both have with the scientific thought
of the age, the differences between us are essential, while
the correspondences are non-essential. And I venture to
think that kinship must be determined by essentials, and
not by non-essentials.*

Joined with the ambiguous use of the phrase "Positive
Philosophy," which has led to a classing with M. Comte
of many men who either ignore or reject his distinctive
principles, there has been one special circumstance that has
tended to originate and maintain this classing in my own
case. The assumption of some relationship between M. Comte
and myself, was unavoidably raised by the title of my first
book—*Social Statics*. When that book was published, I was
unaware that this title had been before used: had I
known the fact, I should certainly have adopted an alternative
title which I had in view.† If, however, instead of the title,

* In his recent work, *Auguste Comte et la Philosophie Positive*, M. Littré,
defending the Comtean classification of the sciences from the criticism I made
upon it in the "Genesis of Science," deals with me wholly as an antagonist.
The chapter he devotes to his reply, opens by placing me in direct antithesis
to the English adherents of Comte, named in the preceding chapter.

† I believed at the time, and have never doubted until now, that the choice
of this title was absolutely independent of its previous use by M. Comte. While
writing these pages, I have found reason to think the contrary. On referring to *Social
Statics*, to see what were my views of social evolution in 1850, when M. Comte
was to me but a name, I met with the following sentence:—"Social philosophy
. may be aptly divided (as political economy has been) into statics and dynamics."
(p. 409). This I remembered to be a reference to a division which I had seen in
the Political Economy of Mr. Mill. But why had I not mentioned Mr. Mill's name?
On referring to the first edition of his work, I found, at the opening of Book iv.,
this sentence:—"The three preceding parts include as detailed a view as the limits
of this treatise permit, of what, by a happy generalization of a mathematical
phrase, has been called the Statics of the subject." Here was the solution of the
question. The division had not been made by Mr. Mill, but by some writer
(on Political Economy I supposed) who was not named by him; and whom I did
not know. It is now manifest, however, that while I supposed I was giving
a more extended use to this division, I was but returning to the original use

the work itself be considered, its irrelation to the philosophy
of M. Comte, becomes abundantly manifest. There is decisive
testimony on this point. In the *North British Review* for
August, 1851, a reviewer of *Social Statics* says—

"The title of this work, however, is a complete misnomer.
According to all analogy, the phrase "Social Statics" should be
used only in some such sense as that in which, as we have already
explained, it is used by Comte, namely as designating a branch of
inquiry whose end it is to ascertain the laws of social equilibrium
or order, as distinct ideally from those of social movement or progress.
Of this Mr. Spencer does not seem to have had the slightest notion,
but to have chosen the name for his work only as a means of indi-
cating vaguely that it proposed to treat of social concerns in a
scientific manner." p. 321.

Respecting M. Comte's application of the words *statics*
and *dynamics* to social phenomena, now that I know what
it is, I will only say that while I perfectly understand how,
by a defensible extension of their mathematical meanings,
the one may be used to indicate social *functions in balance*,
and the other social *functions out of balance*, I am quite at a
loss to understand how the phenomena of *structure* can be
included in the one any more than in the other. But the
two things which here concern me, are, first, to point out that
I had not "the slightest notion" of giving Social Statics the
meaning which M. Comte gave it; and, second, to explain
the meaning which I did give it. The units of any aggregate of matter, are in equilibrium when they severally
act and re-act upon each other on all sides with equal forces.
A state of change among them implies that there are forces
exercised by some that are not counterbalanced by like
forces exercised by others; and a state of rest implies the
absence of such uncounterbalanced forces—implies, if the ·
units are homogeneous, equal distances among them—
implies a maintenance of their respective spheres of molecular

which Mr. Mill had limited to his special topic. Another thing is, I think,
tolerably manifest. As I evidently wished to point out my obligation to some
unknown political economist, whose division I thought I was extending, I should
have named him had I known who he was. And in that case should not have
put this extension of the division as though it were new

motion. Similarly among the units of a society, the fundamental condition to equilibrium, is, that the restraining forces
which the units exercise on each other, shall be balanced.
If the spheres of action of some units are diminished by
extension of the spheres of action of others, there necessarily
results an unbalanced force which tends to produce political
change in the relations of individuals; and the tendency
to change can cease, only when individuals cease to aggress
on each other's spheres of action—only when there is
maintained that law of equal freedom, which it was the
purpose of *Social Statics* to enforce in all its consequences.
Besides this totally-unlike conception of what constitutes
Social Statics, the work to which I applied that title, is
fundamentally at variance with M. Comte's teachings in
almost everything. So far from alleging, as M. Comte does,
that society is to be re-organized by philosophy; it alleges
that society is to be re-organized only by the accumulated
. effects of habit on character. Its aim is not the increase
of authoritative control over citizens, but the decrease of it.
A more pronounced individualism, instead of a more pronounced nationalism, is its ideal. So profoundly is my
political creed at variance with the creed of M. Comte, that,
unless I am misinformed, it has been instanced by a leading
English disciple of M. Comte, as the creed to which he has
the greatest aversion. One point of coincidence, however,
is recognizable. The analogy between an individual organism
and a social organism, which was held by Plato and by
Hobbes, is asserted in *Social Statics*, as it is in the *Sociology*
of M. Comte. Very rightly, M. Comte has made this
analogy the cardinal idea of this division of his philosophy.
In *Social Statics*, the aim of which is essentially ethical,
this analogy is pointed out incidentally, to enforce certain
ethical considerations; and is there obviously suggested
partly by the definition of life which Coleridge derived from
Schelling, and partly by the generalizations of physiologists
there referred to (chap. xxx. §§. 12, 13, 16). Excepting

this incidental agreement, however, the contents of *Social
Statics* are so wholly antagonistic to the philosophy of
M. Comte, that, but for the title, the work would never,
I think, have raised the remembrance of him—unless, indeed,
by the association of opposites.[*]

And now let me point out that which really *has* exercised
a profound influence over my course of thought. The truth
which Harvey's embryological inquiries first dimly indicated,
which was afterwards more clearly perceived by Wolff, and
which was put into a definite shape by Von Baer—the truth
that all organic development is a change from a state of
homogeneity to a state of heterogeneity—this it is from
which very many of the conclusions which I now hold,
have indirectly resulted. In *Social Statics*, there is everywhere manifested a dominant belief in the evolution of man
and of society. There is also manifested the belief that this
evolution is in both cases determined by the incidence of
conditions—the actions of circumstances. And there is
further, in the sections above referred to, a recognition of
the fact that organic and social evolutions, conform to the
same law. Falling amid beliefs in evolutions of various
orders, everywhere determined by natural causes (beliefs again
displayed in the *Theory of Population* and in the *Principles
of Psychology*); the formula of Von Baer acted as an
organizing principle. The extension of it to other kinds
of phenomena than those of individual and social organiza-

[*] Let me add that the conception developed in *Social Statics*, dates back to a
series of letters on the "Proper Sphere of Government," published in the
Nonconformist newspaper, in the latter half of 1842, and republished as a
pamphlet in 1843. In these letters will be found, along with many crude ideas,
the same belief in the conformity of social phenomena to unvariable laws; the
same belief in human progression as determined by such laws; the same belief
in the moral modification of men as caused by social discipline; the same
belief in the tendency of social arrangements "of themselves to assume
a condition of *stable* equilibrium;" the same repudiation of state-control over
various departments of social life; the same limitation of state-action to the
maintenance of equitable relations among citizens. The writing of *Social Statics*
arose from a dissatisfaction with the basis on which the doctrines set forth in those
letters were placed: the second half of that work is an elaboration of these
doctrines; and the first half a statement of the principles from which they are
deducible.

tion, is traceable through successive stages. It may be seen in the last paragraph of an essay on "The Philosophy of Style," published in October, 1852; again in an essay on "Manners and Fashion," published in April, 1854; and then, in a comparatively advanced form, in an essay on "Progress: its Law and Cause," published in April, 1857. Afterwards, there came the recognition of the need for further limitation of this formula; next the inquiry into those general laws of force from which this universal transformation necessarily results; next the deduction of these from the ultimate law of the persistence of force; next the perception that there is everywhere a process of Dissolution complementary to that of Evolution; and, finally, the determination of the conditions (specified in the foregoing essay) under which Evolution and Dissolution respectively occur. The filiation of these results, is, I think, tolerably manifest. The process has been one of continuous development, set up by the addition of Von Baer's law to a number of ideas that were in harmony with it. And I am not conscious of any other influences by which the process has been affected.

It is possible, however, that there may have been influences of which I am not conscious; and my opposition to M. Comte's system may have been one of them. The presentation of antagonistic thoughts, often produces greater definiteness and development of one's own thoughts. It is probable that the doctrines set forth in the essay on "The Genesis of Science," might never have been reached, had not my very decided dissent from M. Comte's conception, led me to work them out; and but for this, I might not have arrived at the classification of the sciences exhibited in the foregoing essay. Very possibly there are other cases in which the stimulus of repugnance to M. Comte's views, may have aided in elaborating my own views; though I cannot call to mind any other cases.

Let it by no means be supposed from all I have said, that I do not regard M. Comte's speculations as of great value.

True or untrue, his system as a whole, has doubtless produced
important and salutary revolutions of thought in many
minds; and will doubtless do so in many more. Doubtless,
too, not a few of those who dissent from his general views,
have been healthfully stimulated by the consideration of them.
The presentation of scientific knowledge and method as a
whole, whether rightly or wrongly co-ordinated, cannot have
failed greatly to widen the conceptions of most of his readers.
And he has done especial service by familiarizing men with
the idea of a social science, based on the other sciences.
Beyond which benefits resulting from the general character
and scope of his philosophy, I believe that there are scattered
through his pages, many large ideas that are valuable not
only as stimuli, but for their actual truth.

It has been by no means an agreeable task to make these
personal explanations; but it has seemed to me a task not to
be avoided. Differing so profoundly as I do from M. Comte
on all fundamental doctrines, save those which we inherit in
common from the past; it has become needful to dissipate
the impression that I agree with him—needful to show that
a large part of what is currently known as "positive
philosophy," is not "positive philosophy" in the sense of
being peculiarly M. Comte's philosophy; and to show that
beyond that portion of the so-called "positive philosophy"
which is not peculiar to him, I dissent from it.

And now at the close, as at the outset, let me express my
great regret that these explanations should have been called
forth by the statements of a critic who has treated me so liber-
ally. Nothing will, I fear, prevent the foregoing pages from
appearing like a very ungracious response to M. Laugel's
sympathetically-written review. I can only hope that the
gravity of the question at issue, in so far as it concerns
myself, may be taken in mitigation, if not as a sufficient
apology.

March 12th, 1864.

APPENDIX.

[*The following chapter was contained in the first edition of*
First Principles. *I omitted it from the re-organized second
edition, because it did not form an essential part of the new
structure. As it is referred to in the foregoing pages, and as
its general argument is germane to the contents of those pages,
I have thought well to append it here. Moreover, though I
hope eventually to incorporate it in that division of the* Principles *of* Sociology *which treats of Intellectual Progress,
yet as it must be long before it can thus re-appear in its permanent place, and as, should I not get so far in the execution
of my undertaking, it may never thus re-appear at all, it seems
proper to make it more accessible than it is at present. The
first and last sections, which served to link it into the argument
of the work to which it originally belonged, are omitted. The
rest has been carefully revised, and in some parts considerably
altered.*]

LAWS IN GENERAL.

The recognition of Law being the recognition of uniformity of relations among phenomena, it follows that the
order in which different groups of phenomena are reduced to
law, must depend on the frequency with which the uniform
relations they severally display are distinctly experienced.
At any given stage of progress, those uniformities will be
best known with which men's minds have been oftenest and
most strongly impressed. In proportion partly to the
number of times a relation has been presented to consciousness (not merely to the senses), and in proportion

ı

partly to the vividness with which the terms of the relation have been cognized, will be the degree in which the constancy of connexion is perceived.

The succession in which relations are generalized being thus determined, there result certain derivative principles to which this succession must more immediately and obviously conform. First is *the directness with which personal welfare is affected.* While, among surrounding things, many do not appreciably influence us in any way, some produce pleasures and some pains, in various degrees; and manifestly, those things whose actions on the organism for good or evil are most decided, will, cæteris paribus, be those whose laws of action are earliest observed. Second comes *the conspicuousness of one or both phenomena between which a relation is to be perceived.* On every side are phenomena so concealed as to be detected only by close observation; others not obtrusive enough to attract notice; others which moderately solicit the attention; others so imposing or vivid as to force themselves on consciousness; and, supposing conditions to be the same, these last will of course be among the first to have their relations generalized. In the third place, we have *the absolute frequency with which the relations occur.* There are coexistences and sequences of all degrees of commonness, from those which are ever present to those which are extremely rare; and manifestly, the rare coexistences and sequences, as well as the sequences which are very long in taking place, will not be reduced to law so soon as those which are familiar and rapid. Fourthly has to be added *the relative frequency of occurrence.* Many events and appearances are limited to certain times or certain places, or both; and, as a relation which does not exist within the environment of an observer cannot be perceived by him, however common it may be elsewhere or in another age, we have to take account of the surrounding physical circum-

stances, as well as of the state of society, of the arts, and of
the sciences—all of which affect the frequency with which
certain groups of facts are observable.   The
fifth corollary to be noticed is, that the succession in
which different classes of relations are reduced to law, depends in part on their *simplicity.* Phenomena presenting
great composition of causes or conditions, have their essential
relations so masked, that it requires accumulated experiences
to impress upon consciousness the true connexions of antecedents and consequents they involve. Hence, other things
equal, the progress of generalization will be from the simple
to the complex ; and this it is which M. Comte has wrongly
asserted to be the sole regulative principle of the progress.   Sixth comes *the degree of abstractness.*
Concrete relations are the earliest acquisitions. Such analyses of them as separate the essential connexions from their
disguising accompaniments, necessarily come later. The
analyses of the connexions, always more or less compound,
into their elements then becomes possible. And so on continually, until the highest and most abstract truths have
been reached.

These, then, are the several derivative principles. The
frequency and vividness with which uniform relations are
repeated in conscious experience, determining the recognition
of their uniformity, and this frequency and vividness depend-
ing on the above conditions, it follows that the order in
which different classes of facts are generalized, must depend
on the extent to which the above conditions are fulfilled in
each class. Let us mark how the facts harmonize with this
conclusion: taking first a few that elucidate the general
truth, and afterwards some that exemplify the special truths
which we here see follow from it.

The relations earliest known as uniformities, are those sub-
sisting between the common properties of matter—tangi-

bility, visibility, cohesion, weight, etc. We have no trace of a time when the resistance offered by an object was regarded as caused by the will of the object; or when the pressure of a body on the hand holding it, was ascribed to the agency of a living being. And accordingly, these are the relations of which we are oftenest conscious; being objectively frequent, conspicuous, simple, concrete, and of immediate personal concern.

Similarly with the ordinary phenomena of motion. The fall of a mass on the withdrawal of its support, is a sequence which directly affects bodily welfare, is conspicuous, simple, concrete, and very often repeated. Hence it is one of the uniformities recognized before the dawn of tradition. We know of no era when movements due to terrestrial gravitation were attributed to volition. Only when the relation is obscured—only, as in the case of an aërolite, where the antecedent of the descent is unperceived, do we find the conception of personal agency. On the other hand, motions of intrinsically the same order as that of a falling stone —those of the heavenly bodies—long remain ungeneralized; and until their uniformity is seen, are construed as results of will. This difference is clearly not dependent on comparative complexity or abstractness; since the motion of a planet in an ellipse, is as simple and concrete a phenomenon as the motion of a projected arrow in a parabola. But the antecedents are not conspicuous; the sequences are of long duration; and they are not often repeated. And that these are the causes of their slow reduction to law, we see in the fact that they are severally generalized in the order of their frequency and conspicuousness—the moon's monthly cycle, the sun's annual change, the periods of the inferior planets, the periods of the superior planets.

While astronomical sequences were still ascribed to volition, certain terrestrial sequences of a different kind, but some of them equally without complication, were interpreted in like manner. The solidification of water at a low temperature, is a phenomenon that is simple, concrete, and of much personal concern. But it is neither so frequent as those which we see are earliest generalized, nor is the presence of the antecedent so manifest. Though in all but tropical climates, mid-winter displays the relation between cold and freezing with tolerable constancy; yet, during the spring and autumn, the occasional appearance of ice in the mornings has no very obvious connexion with coldness of the weather. Sensation being so inaccurate a measure, it is not possible for the savage to experience the definite relation between a temperature of 32° and the congealing of water; and hence the long continued belief in personal agency. Similarly, but still more clearly, with the winds. The absence of regularity and the inconspicuousness of the antecedents, allowed the mythological explanation to survive for a great period.

During the era in which the uniformity of many quite simple inorganic relations was still unrecognized, certain organic relations, intrinsically very complex and special, were generalized. The constant coexistence of feathers and a beak, of four legs with an internal bony framework, are facts which were, and are, familiar to every savage. Did a savage find a bird with teeth, or a mammal clothed with feathers, he would be as much surprised as an instructed naturalist. Now these uniformities of organic structure thus early perceived, are of exactly the same kind as those more numerous ones later established by biology. The constant coexistence of mammary glands with two occipital condyles to the skull, of vertebræ with teeth lodged in sockets, of frontal horns with the habit of rumination, are generalizations as purely empirical as those known to the aboriginal hunter. The botanist cannot in the least understand the complex relation between papilionaceous flowers and seeds borne in flattened pods: he knows these and like connexions simply in the same way that the barbarian knows the connexions between particular leaves and particular kinds of wood. But tho fact that sundry of the uniform relations which chiefly make up the organic sciences, were very early recognized, is due to the high degree of vividness and frequency with which they were presented to consciousness. Though the connexion between the sounds characteristic of a bird, and the possession of edible flesh, is extremely involved; yet the two terms of the relation are conspicuous, often recur in experience, and a knowledge of their connexion has a direct bearing on personal welfare. Meanwhile innumerable relations of the same order, which are displayed with even greater frequency by surrounding plants and animals, remain for thousands of years unrecognised, if they are unobtrusive or of no apparent moment.

When, passing from this primitive stage to a more advanced stage, we trace the discovery of those less familiar uniformities which mainly constitute what is distinguished as Science, we find the succession in which knowledge of them is reached, to bo still determined in the same manner. This will become obvious on contemplating separately the influence of each derivative condition.

How relations that have immediate bearings on the maintenance of life, are, other things equal, fixed in tho mind before those which have no immediate bearings, the history of Science abundantly illustrates. The habits of existing uncivilized races, who fix times by moons and barter so many of one article for so many of another, show us that conceptions of equality and number, which are the germs of mathematical science, were developed under the immediate pressure of personal wants; and it can scarcely be doubted that those laws of numerical relations which are embodied in the rules of arithmetic, were first brought to light through the practice of mercantile exchange. Similarly with geometry. The derivation of the word shows us that it ori-

ginally included only certain methods of partitioning ground
and laying out buildings. The properties of the scales and
the lever, involving the first principle in mechanics, were
early generalized under the stimulus of commercial and
architectural needs. To fix the times of religious festivals
and agricultural operations, were the motives which led to
the establishment of the simpler astronomic periods. Such
small knowledge of chemical relations as was involved in
ancient metallurgy, was manifestly obtained in seeking how
to improve tools and weapons. In the alchemy of later
times, we see how greatly an intense hope of private benefit
contributed to the disclosure of a certain class of uniformities.
Nor is our own age barren of illustrations. "Here," says
Humboldt, when in Guiana, "as in many parts in Europe, the
sciences are thought worthy to occupy the mind, only so far
as they confer some immediate and practical benefit on
society." "How is it possible to believe," said a missionary
to him, "that you have left your country to come to be de-
voured by mosquitoes on this river, and to measure lands
that are not your own." Our coasts furnish like instances.
Every sea-side naturalist knows how great is the contempt
with which fishermen regard the collection of objects for the
microscope or aquarium. Their incredulity as to the possible
value of such things is so great, that they can scarcely be
induced even by bribes to preserve the refuse of their nets.
Nay, we need not go for evidence beyond daily table-talk.
The demand for "practical science"—for a knowledge that
can be brought to bear on the business of life—joined to the
ridicule commonly vented on scientific pursuits having no
obvious uses, suffice to show that the order in which laws
are discovered greatly depends on the directness with which
they affect our welfare.

That, when all other conditions are the same, obtrusive
relations will be generalized before unobtrusive ones, is so
nearly a truism that examples appear almost superfluous. If

it be admitted that by the aboriginal man, as by the child, the co-existent properties of large surrounding objects are noticed before those of minute objects, and that the external relations which bodies present are generalized before their internal relations, it must be admitted that in subsequent stages of progress, the comparative conspicuousness of relations has greatly affected the order in which they were recognized as uniform. Hence it happened that after the establishment of those very manifest sequences constituting a lunation, and those less manifest ones marking a year, and those still less manifest ones marking the planetary periods, astronomy occupied itself with such inconspicuous sequences as those displayed in the repeating cycle of lunar eclipses, and those which suggested the theory of epicycles and eccentrics; while modern astronomy deals with still more inconspicuous sequences, some of which, as the planetary rotations, are nevertheless the simplest which the heavens present. In physics, the early use of canoes implied an empirical knowledge of certain hydrostatic relations that are intrinsically more complex than sundry static relations not empirically known; but these hydrostatic relations were thrust upon observation. Or, if we compare the solution of the problem of specific gravity by Archimedes with the discovery of atmospheric pressure by Torricelli (the two involving mechanical relations of exactly the same kind), we perceive that the much earlier occurrence of the first than the last was determined, neither by a difference in the irbearings on personal welfare, nor by a difference in the frequency with which illustrations of them came under observation, nor by relative simplicity; but by the greater obtrusiveness of the connexion between antecedent and consequent in the one case than in the other. Among miscellaneous illustrations, it may be pointed out that the connexions between lightning and thunder, and between rain and clouds, were recognized long before others of the same order, simply because they thrust themselves on the attention. Or the long-delayed discovery of the microscopic forms of life, with all the phenomena they present, may bo named as very clearly showing how certain groups of relations not ordinarily perceptible, though in other respects like long-familiar relations, have to wait until changed conditions render them perceptible. But, without further details, it needs only to consider the inquiries which now occupy the electrician, the chemist, the physiologist, to see that science has advanced, and is advancing, from the more conspicuous phenomena to the less conspicuous ones.

How the degree of absolute frequency of a relation affects the recognition of its uniformity, we see in contrasting certain biological facts. The connexion between death and bodily injury, constantly displayed not only in men but in all inferior creatures, was known as an instance of natural causation while yet deaths from diseases were thought supernatural. Among diseases themselves, it is observable that unusual ones were regarded as of demoniacal origin during ages when the more frequent were ascribed to ordinary causes: a truth paralleled among our own peasantry, who by the use of charms show a lingering superstition with respect to rare disorders, which they do not show with respect to common ones, such as colds. Passing to physical illustrations, we may note that within the historic period whirlpools were accounted for by the agency of water-spirits; but wo do not find that within the same period the disappearance of water on exposure either to the sun or to artificial heat was interpreted in an analogous way: though a more marvellous occurrence, and a much more complex one, its great frequency led to the early recognition of it as a natural uniformity. Rainbows and comets do not differ much in conspicuousness, and a rainbow is intrinsically the more involved phenomenon; but chiefly because of their far greater commonness, rainbows were perceived to have a direct dependance on sun and rain while yet comets were regarded as signs of divine wrath.

That races living inland must long have remained ignorant of the daily and monthly sequences of the tides, and that tropical races could not early have comprehended the phenomena of northern winters, are extreme illustrations of the influence which relative frequency has on the recognition of uniformities. Animals which, where they are indigenous, call forth no surprise by their structures or habits, because these are so familiar, when taken to countries where they have never been seen, are looked at with an astonishment approaching to awe—are even thought supernatural: a fact which will suggest numerous others that show how the localization of phenomena in part controls the order in which they are reduced to law. Not only however does their localization in space affect the progression, but also their localization in time. Facts which are rarely if ever manifested in one era, are rendered very frequent in another, simply through the changes wrought by civilization. The lever, of which the properties are illustrated in the use of sticks and weapons, is vaguely understood by every savage—on applying it in a certain way he rightly anticipates certain effects; but the wheel-and-axle, pulley, and screw, cannot have their powers either empirically or rationally known till the advance of the arts has more or less familiarized them. Through those various means of exploration which we have inherited and added to, we have become acquainted with a vast range of chemical relations that were relatively non-existent to the primitive man. To highly-developed industries we owe both the substances and the appliances that have disclosed to us countless uniformities which our ancestors had no opportunity of seeing. These and like instances that will occur to the reader, show that the accumulated materials, and processes, and products, which characterize the environments of complex societies, greatly increase the accessibility of various

59

classes of relations; and by so multiplying the experiences
of them, or making them relatively frequent, facilitate their
generalization. Moreover, various classes of phenomena
presented by society itself, as for instance those which
political economy formulates, become relatively frequent, and
therefore recognizable, in advanced social states; while in
less advanced ones they are either too rarely displayed to
have their relations perceived, or, as in the least advanced
ones, are not displayed at all.

That, where no other circumstances interfere, the order in
which different uniformities are established varies as their
complexity, is manifest. The geometry of straight lines was
understood before the geometry of curved lines; the properties of the circle before the properties of the ellipse, parabola,
and hyperbola; and the equations of curves of single curvature were ascertained before those of curves of double
curvature. Plane trigonometry comes in order of time and
simplicity before spherical trigonometry; and the mensuration of plane surfaces and solids before the mensuration of
curved surfaces and solids. Similarly with mechanics: the
laws of simple motion were generalized before those of compound motion; and those of rectilinear motion before those
of curvilinear motion. The properties of equal-armed levers
or scales, were understood before those of levers with un-
equal arms; and the law of the inclined plane was formulated
earlier than that of the screw, which involves it. In chemistry, the progress has been from the simple inorganic compounds to the more involved or organic compounds. And
where, as in the higher sciences, the conditions of the exploration are more complicated, we still may clearly trace
relative complexity as determining the order of discovery
where other things are equal.

The progression from concrete relations to abstract ones,
and from the less abstract to the more abstract, is equally
obvious. Numeration, which in its primary form concerned

itself only with groups of actual objects, came earlier than simple arithmetic; the rules of which deal with numbers apart from objects. Arithmetic, limited in its sphere to concrete numerical relations, is alike earlier and less abstract than Algebra, which deals with the relations of these relations. And in like manner, the Calculus of Operations comes after Algebra, both in order of evolution and in order of abstractness. In Mechanics, the more concrete relations of forces exhibited in the lever, inclined plane, etc., were understood before the more abstract relations expressed in the laws of resolution and composition of forces; and later than the three abstract laws of motion as formulated by Newton came the still more abstract law of inertia. Similarly with Physics and Chemistry, there has been an advance from truths entangled in all the specialities of particular facts and particular classes of facts, to truths disentangled from the disguising incidents under which they are manifested—to truths of a higher abstractness.

Brief and rude as is this sketch of a mental development that has been long and complicated, I venture to think it shows inductively what was deductively inferred, that the order in which separate groups of uniformities are recognized, depends not on one circumstance but on several circumstances. The various classes of relations are generalized in a certain succession, not solely because of one particular kind of difference in their natures; but also because they are variously placed in time and in space, variously open to observation, and variously related to our own constitutions: our perception of them being influenced by all these conditions in endless combinations. The comparative degrees of importance, of obtrusiveness, of absolute frequency, of relative frequency, of simplicity, of concreteness, are every one of them factors; and from their unions in proportions that are never twice alike, there results a highly complex process of mental evolution. But while it is thus manifest

that the proximate causes of the succession in which relations
are reduced to law, are numerous and involved; it is also
manifest that there is one ultimate cause to which these
proximate causes are subordinate. As the several circumstances that determine the early or late recognition of uniformities are circumstances that determine the number and
strength of the impressions which these uniformities make
on the mind, it follows that the progression conforms to a
certain fundamental principle of psychology. We see *à*
posteriori, what we concluded *à priori*, that the order in which
relations are generalized, depends on the frequency and
impressiveness with which they are repeated in conscious
experience.

Having roughly analyzed the progress of the past, let
us take advantage of the light thus thrown on the present,
and consider what is implied respecting the future.

Note first that the likelihood of the universality of Law
has been ever growing greater. Out of the countless coexistences and sequences with which mankind are environed,
they have been continually transferring some from the group
whose order was supposed to be arbitrary, to the group
whose order is known to be uniform. And manifestly, as
fast as the relations that are unreduced to law become
fewer, the probability that among them there are some that
do not conform to law, becomes less. To put the argument
numerically—It is clear that when out of surrounding phenomena a hundred of several kinds have been found to occur
in constant connexions, there arises a slight presumption that
all phenomena occur in constant connexions. When uniformity has been established in a thousand cases, more varied
in their kinds, the presumption gains strength. And when
the known cases of uniformity amount to myriads, including
many of each variety, it becomes an ordinary induction that
uniformity exists everywhere.

Silently and insensibly their experiences have been pressing men on towards the conclusion thus drawn. Not out of a conscious regard for those reasons, but from a habit of thought which these reasons formulate and justify, all minds have been advancing towards a belief in the constancy of surrounding coexistences and sequences. Familiarity with concrete uniformities has generated the abstract conception of uniformity—the idea of *Law*; and this idea has been in successive generations slowly gaining fixity and clearness. Especially has it been thus among those whose knowledge of natural phenomena is the most extensive—men of science. The mathematician, the physicist, the astronomer, the chemist, severally acquainted with the vast accumulations of uniformities established by their predecessors, and themselves daily adding new ones as well as verifying the old, acquire a far stronger faith in law than is ordinarily possessed. With them this faith, ceasing to be merely passive, becomes an active stimulus to inquiry. Wherever there exist phenomena of which the dependence is not yet ascertained, these most cultivated intellects, impelled by the conviction that here too there is some invariable connexion, proceed to observe, compare, and experiment; and when they discover the law to which the phenomena conform, as they eventually do, their general belief in the universality of law is further strengthened. So overwhelming is the evidence, and such the effect of this discipline, that to the advanced student of nature, the proposition that there are lawless phenomena has become not only incredible but almost inconceivable.

This habitual recognition of law which already distinguishes modern thought from ancient thought, must spread among men at large. The fulfilment of predictions made possible by every new step, and the further command gained of nature's forces, prove to the uninitiated the validity of scientific generalizations and the doctrine they illustrate. Widening education is daily diffusing among the mass of

men that knowledge of these generalizations which has been
hitherto confined to the few. And as fast as this diffusion
goes on, must the belief of the scientific become the belief of
the world at large.

That law is universal, will become an irresistible conclusion when it is perceived that *the progress in the discovery of laws itself conforms to law*; and when this perception makes it clear why certain groups of phenomena have
been reduced to law, while other groups are still unreduced.
When it is seen that the order in which uniformities are
recognized, must depend upon the frequency and vividness
with which they are repeated in conscious experience; when
it is seen that, as a matter of fact, the most common, important, conspicuous, concrete, and simple, uniformities were the
earliest recognized, because they were experienced oftenest
and most distinctly; it will by implication be seen that long
after the great mass of phenomena have been generalized,
there must remain phenomena which, from their rareness,
or unobtrusiveness, or seeming unimportance, or complexity,
or abstractness, are still ungeneralized. Thus will be
furnished a solution to a difficulty sometimes raised. When
it is asked why the universality of law is not already fully
established, there will be the answer that the directions in
which it is not yet established are those in which its establishment must necessarily be latest. That state of things
which is inferable beforehand, is just the state which we find
to exist. If such coexistences and sequences as those of
Biology and Sociology are not yet reduced to law, the presumption is not that they are irreducible to law, but that their
laws elude our present means of analysis. Having long ago
proved uniformity throughout all the lower classes of relations, and having been step by step proving uniformity
throughout classes of relations successively higher and higher,
if we have not yet succeeded with the highest classes, it may

be fairly concluded that our powers are at fault, rather
than that the uniformity does not exist. And unless we
make the absurd assumption that the process of generalization, now going on with unexampled rapidity, has reached
its limit, and will suddenly cease, we must infer that ul-
timately mankind will discover a constant order of manifestation even in the most involved and obscure phenomena.

www.ingramcontent.com/pod-product-compliance
Lightning Source LLC
Chambersburg PA
CBHW021947190326
41519CB00009B/1169